Springer Series in Optical Sciences

Volume 254

Springer Series in Optical Sciences is led by Editor-in-Chief William T. Rhodes, Florida Atlantic University, USA, and provides an expanding selection of research monographs in all major areas of optics:

- lasers and quantum optics
- ultrafast phenomena
- optical spectroscopy techniques
- optoelectronics
- information optics
- applied laser technology
- industrial applications and
- synchrotron radiation and X-ray optics
- other topics of contemporary interest.

With this broad coverage of topics the series is useful to research scientists and engineers who need up-to-date reference books.

Jun Shibayama

FDTD Analysis of Plasmonic Devices

Springer

Jun Shibayama
Department of Electrical and Electronic Engineering
Hosei University
Koganei, Tokyo, Japan

ISSN 0342-4111 ISSN 1556-1534 (electronic)
Springer Series in Optical Sciences
ISBN 978-981-96-8129-7 ISBN 978-981-96-8130-3 (eBook)
https://doi.org/10.1007/978-981-96-8130-3

This Springer imprint is published by the registered company Springer Nature Singapore Pte Ltd.
The registered company address is: 152 Beach Road, #21-01/04 Gateway East, Singapore 189721, Singapore

Preface

Surface plasmon is a type of surface wave generated by the coupling of light and electrons at the metal-dielectric interface. Since surface plasmon is confined to a region smaller than the wavelength, it has been actively studied for miniaturization of optical circuits. The characteristics of surface plasmon are quite sensitive to its environment, so it has also been put into practical use as a surface plasmon resonance sensor.

This book is a compilation of the author's research results on the analysis of plasmonic devices using the finite-difference time-domain (FDTD) method, which is written for engineers, researchers, and graduate students who want to analyze plasmonic devices. In addition to the optical band, many pages are devoted to plasmonic devices in the terahertz band as a new horizon for the use of surface plasmons. First, the fundamentals of the FDTD method are presented in Chaps. 1 and 2. Chapters 3–11 present the results of specific device analyses. Where necessary, the formulation of the method is also explained. The book is organized as follows.

Chapter 1 presents the fundamentals of the finite-difference time-domain (FDTD) method. The detailed formulation of the traditional explicit scheme is given and its stability condition is presented. Some implicit formulations, which eliminate or relax this condition, are also provided for problems where the sampling width should be chosen extremely small. The concept of the convolutional perfectly matched layer (CPML) is given as a robust absorbing boundary condition for the practical simulations.

Chapter 2 deals with the frequency-dependent formulation of the FDTD method for dispersive media. The three widely used techniques are given: the Auxiliary Differential Equation (ADE), the Recursive Convolution (RC) and the Z-Transformation (ZT) techniques. Among them, the trapezoidal RC (TRC) is a simple, accurate technique that we mainly use for the analyses in this book. For dispersive media, the Drude, Drude-Lorentz, and Drude-Critical Points (D-CP) models are discussed. The implicit local one-dimensional (LOD) formulation is also provided, which is quite effective in reducing the computational time for the analysis of plasmonic devices.

Chapter 3 provides an analysis of some plasmonic devices consisting of a metalinsulator-metal (MIM) waveguide and a metal rod. Band-stop and band-pass filters made of MIM waveguides are analyzed. In particular, the two-dimensional (2-D) and three-dimensional (3-D) results are compared and discussed. It is shown that the performance of the 2-D model is difficult to reproduce by simply adopting the 3-D model. A reflection filter made of a metal rod grating is also analyzed using the body-of-revolution (BOR) FDTD method based on the LOD scheme.

Chapter 4 offers a demultiplexer for separating wavelengths of 1.31 and 1.55 μm based on 3-D MIM waveguides. To achieve a narrower bandwidth for the slit-type filter, the metal thickness is optimized and a dielectric material is inserted into the slit section. To achieve a separation of more than 20 dB for both wavelengths, two stub-type filters are also utilized in the demultiplexer. The analysis is performed using a GPU-accelerated FDTD method based on the TRC technique.

Chapter 5 contains an analysis of a novel surface plasmon resonance (SPR) waveguide sensor with a thin InSb layer in the sensing section in the THz region. The interesting point is to use a semiconductor material InSb to have the SPR in the THz region. Numerical results for the 2-D sensor show that it can detect temperature variations of water from 280 to 320 K and mixtures of organic solvents and water. The analyte container is added to the sensor and the sensor structure is optimized. A 3-D sensor is also designed and analyzed to demonstrate its performance for practical use.

Chapter 6 describes numerical studies of Kretschmann- and Otto-type SPR waveguide sensors at THz frequencies. The motivation is to improve the sensitivity to temperature variations. The sensor is designed using the Fresnel equation and the frequency response is analyzed using the TRC-FDTD method. It is found for the Kretschmann-type sensor that a high sensitivity of 2.63×10^{-2} THz/K, more than three times that of the conventional SPR waveguide sensor in Chap. 5, is successfully obtained. The Otto-type waveguide sensor is shown to maintain the same sensitivity as that of the Kretschmann type, while providing steep and stable SPR responses.

Chapter 7 discusses a TM-pass/TE-stop THz waveguide polarizer, since only the TM waves are needed as an input field for the SPR sensor. For an efficient analysis, a hybrid-implicit-explicit FDTD (HIE-FDTD) method is developed in which the implicit scheme is introduced only in the specific direction where small sampling widths are required. The 2-D and 3-D polarizers with InAs layer are shown to provide an extinction ratio of more than 20 dB, in which the computational time can be significantly reduced. A 3-D polarizer and a waveguide with InAs sidewalls are also investigated. In particular, a waveguide with air buffers and InAs sidewalls is found to act as a TE/TM pass waveguide when the buffer is filled with air, and as a TM-pass/TE-stop polarizer when the buffer is filled with water, showing the possibility of tunable devices.

Chapter 8 involves a grating coupler consisting of periodic arrays of InSb-coated dielectric cylinders on a substrate at THz frequencies. The transmission characteristics of an infinite periodic array are studied not only at normal incidence but also at oblique incidence. It is shown that the incident field is coupled to the substrate due to the guided-mode resonance (GMR), indicating the practical application of a grating coupler. The sensor application is also discussed with emphasis on SPR excitation. The field radiation from a finite array of gratings is explained using a vector diagram. A 3-D analysis of an InSb sphere array on a dielectric substrate is also performed and discussed for application to grating couplers and sensors.

Chapter 9 considers a novel method for splitting THz surface waves based on the propagation characteristics of spoof surface plasmon plaritons. Metal disk splitters with a central hole and radially arranged gratings are proposed and analyzed using the FDTD method in cylindrical coordinates. The splitting behavior is demonstrated by numerical and experimental results. To analyze cylindrical structures efficiently, the HIE-FDTD method is also developed in the cylindrical coordinate systems, where the implicit scheme is introduced only in the circumferential direction. The computation is successfully reduced to less than 20% of the explicit method, yielding almost the same numerical results.

Chapter 10 formulates an extended FDTD algorithm for the analysis of arbitrarily shaped dielectric and dispersive media. For dielectric and dispersive media, the contour-path effective-permittivity (CP-EP) and dispersive contour-path (DCP) algorithms are extended to 3-D problems, respectively. Comparison with results from Mie theory and the conventional staircase approximation shows that the CP-EP and DCP algorithms yield faster numerical convergence with respect to the size of the sampling widths for the problems with arbitrarily shaped structures.

Chapter 11 examines the properties of a plasmonic grating coupler. In particular, the technique for unidirectional excitation of normally incident light on the metal substrate is studied. A trapezoidal dielectric grating coupler for unidirectional propagation at normal incidence is proposed. It is found that the coupling light in the grating propagates unidirectionally on the metal surface even at normal incidence. The effect of adding an antireflection coating on the coupling characteristics is also investigated.

It is hoped that the device analysis presented in this book will provide the reader with useful knowledge for practical plasmonic device applications not only in the optical but also in the THz waveband.

Finally, I would like to express my sincere gratitude to the Japan Society for the Promotion of Science (JSPS) for their financial support through the Grant-in-Aid for Scientific Research (KAKENHI). Part of the research presented in this book was conducted under the support of KAKENHI Grant Number 21760266, 23560415, 15K06035 and 19K04537. This funding played a crucial role in advancing the analysis of plasmonic devices using the FDTD method.

Koganei, Japan Jun Shibayama

Acknowledgements I would like to thank Prof. K. Kobayashi (Chuo University) for giving me the opportunity to publish this book. Thanks to Dr. A. Tokuno and Ms. K. Palanisamy (Springer) for encouraging me to write this book. Thanks are also due to Professor Emeritus J. Yamauchi (Hosei University) and Professor Emeritus H. Nakano (Hosei University) for their constant support of my research work.

Special thanks are expressed to the members of Shibayama Laboratory, Hosei University, who were involved in the realization of this book: M. Muraki (Chaps. 1 and 2), R. Takahashi (Chap. 2), A. Nomura (Chap. 2), R. Ando (Chap. 2), K. Watanabe (Chap. 2), T. Hirano (Chap. 2), B. Murakami (Chap. 3), T. Oikawa (Chap. 3), Y. Wada (Chap. 3), H. Kawai (Chap. 4), K. Shimizu (Chap. 5), K. Yoshihara (Chap. 5), H. Tanaka (Chap. 5), K. Mitsutake (Chap. 6), S. Gomi (Chap. 7), R. Otsuka (Chap. 7), H. Miyao (Chap. 7), S. Takahashi (Chap. 8), T. Kuroda (Chap. 8), R. Ochiai (Chap. 9), D. Kusunoki (Chap. 9), M. Ito (Chap. 9), T. Hara (Chap. 9), K. Suzuki (Chap. 10), T. Iwamoto (Chap. 10), K. Takeya (Chap. 10) and H. Nakasaka (Chap. 11).

Finally, I would like to express my deepest gratitude to my wife, Rikako, for her constant support and encouragement. This book could not have been completed without her unwavering support and dedication.

Koganei, Japan — Jun Shibayama

Competing Interests The author has no competing interests to declare that are relevant to the content of this manuscript.

Contents

Acronyms

2-D	Two-dimensional
3-D	Three-dimensional
ACC	Accelerators
ADE	Auxiliary differential equation
ADI	Alternating-direction implicit
AR	Anti-reflection
BOR	Body-of-revolution
BPM	Beam-propagation method
CFL	Courant-Friedrichs-Lewy
CFLN	Courant-Friedrichs-Lewy number
CN	Crank-Nicolson
COP	Cycloolefin polymer
CP-EP	Contour-path effective-permittivity
CPML	Convolutional perfectly matched layer
CPU	Central processing unit
DCP	Dispersive contour pass
D-CP	Drude-Critical points
DNA	Deoxyribonucleic acid
EM	Electromagnetic
FD-FDTD	Frequency-dependent finite-difference time-domain
FDTD	Finite-difference time-domain
GMR	Guided-mode resonance
GPU	Graphics processing unit
HIE	Hybrid implicit-explicit
InAs	Indium arsenide
InSb	Indium antimonide
LOD	Locally one-dimensional
MIM	Metal-insulator-metal
MP	Multiprocessing
PE	Polyethylene
PEC	Perfect electric conductor

PLRC	Piecewise linear recursive convolution
PML	Perfectly matched layer
RC	Recursive convolution
RIU	Refractive index unit
SCS	Scattering cross section
SPP	Surface plasmon polariton
SPR	Surface plasmon resonance
TE	Transverse electric
TM	Transverse magnetic
TRC	Trapezoidal recursive convolution
YM-BPM	Yee-mesh-based beam-propagation method
ZT	Z-transform

Chapter 1
FDTD Method

Abstract This chapter presents the fundamentals of the finite-difference time-domain (FDTD) method, in which the formulation is provided for the case of linear, isotropic and lossless media. In addition to the traditional explicit FDTD method, also given are the formulations of the implicit FDTD methods based on the Crank–Nicolson (CN), alternating-direction implicit (ADI), and locally one-dimensional (LOD) schemes. Some alternative techniques for implementing implicit FDTD method are provided. Furthermore, the hybrid implicit-explicit (HIE) FDTD method is explained, which is also attractive for the problem where small meshes are required only in the specific direction. The application of the absorbing boundary condition is quite important for the practical use of the FDTD method. The formulation of the convolutional perfect matched layer (CPML) is given, which is used throughout this book.

1.1 Introduction

The finite-difference time-domain (FDTD) method has been used to analyze a wide variety of electromagnetic (EM) problems [1–3]. Nowadays, many commercial software packages are available, so we can easily carry out the FDTD analysis of problems interested. The formulation of the FDTD method is easy only with the knowledge of the finite-difference scheme. However, the implementation is sometime difficult due to the introduction of the absorbing boundary condition, field excitation and visualization of the field calculated. First, we present the formulation of the traditional explicit FDTD method [1], paying attention to the position of electromagnetic components in time and space. It is pointed out that the Courant-Friedrichs-Lewy (CFL) condition should be satisfied for the explicit FDTD method. An implicit FDTD method is then described to eliminate the CFL condition. The Crank–Nicolson (CN)-based FDTD method [4–6] is unconditionally stable and highly accurate but is time-consuming method due to the solution of widely-banded matrix. To efficiently perform the implicit FDTD calculations, operator splitting techniques, such as the alternating-direction implicit (ADI) [7–14] and locally one-dimensional (LOD) [15–21] schemes, are introduced into the implicit FDTD method.

J. Shibayama, *FDTD Analysis of Plasmonic Devices*, Springer Series in Optical Sciences 254, https://doi.org/10.1007/978-981-96-8130-3_1

The fundamental scheme [22] is also effective in implementing the implicit FDTD method, since it provides derivative-free calculations in the right-hand side of the resultant equations. The formulation of the hybrid-implicit-explicit (HIE) FDTD method is presented for the problems where fine meshes are required only for a specific direction, which is solved implicitly [5, 23–27, 29]. Finally, the convolutional perfectly matched layer is formulated for absorbing outgoing waves at the edge of the computational windows [30].

1.2 Explicit Technique

Maxwell's equations for linear, isotropic, and lossless media are given as

$$\nabla \times \boldsymbol{E} = -\mu \frac{\partial \boldsymbol{H}}{\partial t} \tag{1.1}$$

$$\nabla \times \boldsymbol{H} = \varepsilon \frac{\partial \boldsymbol{E}}{\partial t} \tag{1.2}$$

where $\boldsymbol{E}$ and $\boldsymbol{H}$ are the electric and magnetic vectors, respectively, μ is the permeability, and ε is permittivity. From (1.1) and (1.2), we obtain

$$\mu \frac{\partial H_x}{\partial t} = \frac{\partial E_y}{\partial z} - \frac{\partial E_z}{\partial y} \tag{1.3}$$

$$\mu \frac{\partial H_y}{\partial t} = \frac{\partial E_z}{\partial x} - \frac{\partial E_x}{\partial z} \tag{1.4}$$

$$\mu \frac{\partial H_z}{\partial t} = \frac{\partial E_x}{\partial y} - \frac{\partial E_y}{\partial x} \tag{1.5}$$

$$\varepsilon \frac{\partial E_x}{\partial t} = \frac{\partial H_z}{\partial y} - \frac{\partial H_y}{\partial z} \tag{1.6}$$

$$\varepsilon \frac{\partial E_y}{\partial t} = \frac{\partial H_x}{\partial z} - \frac{\partial H_z}{\partial x} \tag{1.7}$$

$$\varepsilon \frac{\partial E_z}{\partial t} = \frac{\partial H_y}{\partial x} - \frac{\partial H_x}{\partial y} \tag{1.8}$$

Now, we discretize (1.3)–(1.8) using the arrangement of the EM fields shown in Fig. 1.1, in which (a) and (b) illustrate the EM fields in space and time, respectively. From Fig. 1.1a, known as Yee's mesh [1], it can be seen that the electric and magnetic fields are spatially shifted by half a cell. The side lengths in the x, y and z directions are represented by Δx, Δy and Δz, respectively. The pace point is expressed as $x = i\Delta x$, $y = j\Delta y$ and $z = k\Delta z$, in which i, j and k are integers. From Fig. 1.1b, the electric and magnetic fields are shifted by half a cell in time. As a result, the electric and magnetic fields are shifted by half a cell in both time and space for the FDTD formulation.

Fig. 1.1 Arrangement of electromagnetic fields

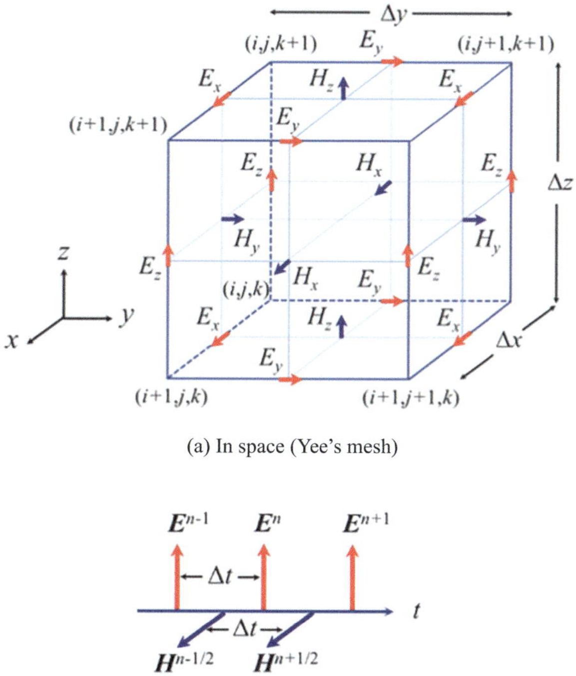

(a) In space (Yee's mesh)

(b) In time

As an example, we derive the finite-difference equation of (1.3). When differentiating the left-hand side of (1.3), the center of the time derivative is set to n. Therefore, we differentiate H_x using the values at $n + 1/2$ and $n - 1/2$ as

$$\frac{\partial H^{n}_{x,i,j+1/2,k+1/2}}{\partial t} \simeq \frac{H^{n+1/2}_{x,i,j+1/2,k+1/2} - H^{n-1/2}_{x,i,j+1/2,k-1/2}}{\Delta t} \tag{1.9}$$

The position of H_x in space is at $i,\ j + 1/2,\ k + 1/2$ in Fig. 1.1a, which is denoted by the subscript of H_x.

Then, we differentiate the right-hand side of (1.3). Since the center of the time derivative is at n, we use Yee's mesh shown in Fig. 1.1a at this time. For the first term in the right-hand side of (1.3), the center of the space derivative is at $i,\ j + 1/2,\ k + 1/2$ of H_x in (1.9). Therefore, across this H_x, we differentiate E_y using the field at $i,\ j + 1/2,\ k + 1$ and $i,\ j + 1/2,\ k$ as

$$\frac{\partial E^{n}_{y,i,j+1/2,k+1/2}}{\partial z} \simeq \frac{E^{n}_{y,i,j+1/2,k+1} - E^{n}_{y,i,j+1/2,k}}{\Delta z} \tag{1.10}$$

Similarly, the second term in the right-hand side of (1.13) is differentiated as

$$\frac{\partial E^n_{z,i,j+1/2,k+1/2}}{\partial y} \simeq \frac{E^n_{z,i,j+1,k+1/2} - E^n_{y,i,j,k+1/2}}{\Delta y} \tag{1.11}$$

Substituting (1.9)–(1.11) into (1.3) and rearranging it, we derive the following finite-difference equation to be solved:

$$\begin{aligned} H^{n+1/2}_{x,i,j+1/2,k+1/2} &= H^{n-1/2}_{x,i,j+1/2,k-1/2} \\ &+ \frac{\Delta t}{\mu \Delta z}\left(E^n_{y,i,j+1/2,k+1} - E^n_{y,i,j+1/2,k}\right) \\ &- \frac{\Delta t}{\mu \Delta y}\left(E^n_{z,i,j+1,k+1/2} - E^n_{y,i,j,k+1/2}\right) \end{aligned} \tag{1.12}$$

Using (1.12), we can obtain the H_x field at $n + 1/2$ with the given initial values at n and $n - 1/2$. The other two components H_y and H_z are similarly derived and calculated. With respect to the electric field, the finite-difference equations are derived from (1.6)–(1.8) and its solution is calculated with the initial field E^n and the calculated field $H^{n+1/2}$. As can be seen, the FDTD method is a simple algorithm, in which the fields at n and $n - 1/2$ are given as the initial values, the entire domain is updated every time step Δt, and the calculation proceeds until the desired time.

The above formulation is based on an explicit scheme, i.e., we obtain the solutions explicitly without solving any matrix. However, at the cost of this, the upper limit of the time step Δt is restricted by the well-known Courant-Friedrichs-Lewy (CFL) condition as

$$v\Delta t \leq \frac{1}{\sqrt{\left(\frac{1}{\Delta x}\right)^2 + \left(\frac{1}{\Delta y}\right)^2 + \left(\frac{1}{\Delta z}\right)^2}} \tag{1.13}$$

where v is the maximum speed of the EM wave in the calculation region. As seen, the upper limit of Δt is determined with the spatial sampling widths. Note, for the problems with small spatial sampling widths, that the time step becomes quire small resulting in a long computation time. To remove this CFL condition, we can resort to some implicit FDTD methods presented in the next section.

1.3 Implicit Technique

1.3.1 Crank–Nicolson Scheme

The CN scheme has been widely used to discretize a parabolic partial differential equation. The advantage of the CN scheme is that the calculation is unconditionally stable, i.e., a large time step beyond the upper limit determined by the CFL condition can be used as long as the calculation accuracy is maintained. Here, we review the formulation of the FDTD method based on the CN scheme [4].

Maxwell's equation is expressed in the matrix form as

$$\frac{\partial \boldsymbol{\phi}}{\partial t} = [\mathcal{A}]\boldsymbol{\phi} \tag{1.14}$$

where $\boldsymbol{\phi} = [E_x, E_y, E_z, H_x, H_y, H_z]^{\mathsf{T}}$ and

$$[\mathcal{A}] = \begin{bmatrix} 0 & 0 & 0 & 0 & -\frac{\partial}{\varepsilon\partial z} & \frac{\partial}{\varepsilon\partial y} \\ 0 & 0 & 0 & \frac{\partial}{\varepsilon\partial z} & 0 & -\frac{\partial}{\varepsilon\partial x} \\ 0 & 0 & 0 & -\frac{\partial}{\varepsilon\partial y} & \frac{\partial}{\varepsilon\partial x} & 0 \\ 0 & \frac{\partial}{\mu\partial z} & -\frac{\partial}{\mu\partial y} & 0 & 0 & 0 \\ -\frac{\partial}{\mu\partial z} & 0 & \frac{\partial}{\mu\partial x} & 0 & 0 & 0 \\ \frac{\partial}{\mu\partial y} & -\frac{\partial}{\mu\partial x} & 0 & 0 & 0 & 0 \end{bmatrix}$$

in which T is the transpose. Here, we apply the CN scheme to (1.14) as

$$\frac{\boldsymbol{\phi}^{n+1} - \boldsymbol{\phi}^{n}}{\Delta t} = [\mathcal{A}]\frac{\boldsymbol{\phi}^{n+1} + \boldsymbol{\phi}^{n}}{2} \tag{1.15}$$

Note, for the implicit FDTD method, that the half time step offset is not usually used for stability. Equation (1.15) is rewritten as follows:

$$\left([I] - \frac{\Delta t}{2}[\mathcal{A}]\right)\boldsymbol{\phi}^{n+1} = \left([I] + \frac{\Delta t}{2}[\mathcal{A}]\right)\boldsymbol{\phi}^{n} \tag{1.16}$$

where $[I]$ is the identity matrix. Note that the CN-FDTD method has second-order accuracy both in time and space. From (1.16), we obtain six equations to be solved. As shown in [4], the resultant implicit equations become a block tridiagonal, the solution of which is quite time-consuming. To efficiently solve the CN-based FDTD equation, some approximate techniques have been proposed [5, 6].

1.3.2 Alternating-Direction Implicit Scheme

Instead of directly calculating the CN scheme, we often resort to operator splitting techniques such as the ADI scheme [7–9]. Here, we review the ADI-FDTD method. The matrix $[\mathcal{A}]$ is decomposed as

$$[\mathcal{A}] = \begin{bmatrix} 0 & 0 & 0 & 0 & 0 & \frac{\partial}{\varepsilon \partial y} \\ 0 & 0 & 0 & \frac{\partial}{\varepsilon \partial z} & 0 & 0 \\ 0 & 0 & 0 & 0 & \frac{\partial}{\varepsilon \partial x} & 0 \\ 0 & \frac{\partial}{\mu \partial z} & 0 & 0 & 0 & 0 \\ 0 & 0 & \frac{\partial}{\mu \partial x} & 0 & 0 & 0 \\ \frac{\partial}{\mu \partial y} & 0 & 0 & 0 & 0 & 0 \end{bmatrix} + \begin{bmatrix} 0 & 0 & 0 & 0 & -\frac{\partial}{\varepsilon \partial z} & 0 \\ 0 & 0 & 0 & 0 & 0 & -\frac{\partial}{\varepsilon \partial x} \\ 0 & 0 & 0 & -\frac{\partial}{\varepsilon \partial y} & 0 & 0 \\ 0 & 0 & -\frac{\partial}{\mu \partial y} & 0 & 0 & 0 \\ -\frac{\partial}{\mu \partial z} & 0 & 0 & 0 & 0 & 0 \\ 0 & -\frac{\partial}{\mu \partial x} & 0 & 0 & 0 & 0 \end{bmatrix}$$

$$= [A] + [B] \tag{1.17}$$

Equation (1.16) can be rewritten with (1.17) as

$$\left([I] - \frac{\Delta t}{2}([A] + [B])\right) \boldsymbol{\phi}^{n+1} = \left([I] + \frac{\Delta t}{2}([A] + [B])\right) \boldsymbol{\phi}^{n}$$

We introduce operator splitting as

$$\left([I] - \frac{\Delta t}{2}[A]\right)\left([I] - \frac{\Delta t}{2}[B]\right) \boldsymbol{\phi}^{n+1} = \left([I] + \frac{\Delta t}{2}[A]\right)\left([I] + \frac{\Delta t}{2}[B]\right) \boldsymbol{\phi}^{n}$$

which is rewritten as

$$\boldsymbol{\phi}^{n+1} = \frac{\left([I] + \frac{\Delta t}{2}[A]\right)\left([I] + \frac{\Delta t}{2}[B]\right)}{\left([I] - \frac{\Delta t}{2}[A]\right)\left([I] - \frac{\Delta t}{2}[B]\right)} \boldsymbol{\phi}^{n} \tag{1.18}$$

Here, we apply the ADI scheme to (1.18), leading to

$$\boldsymbol{\phi}^{n+1/2} = \frac{[I] + \frac{\Delta t}{2}[A]}{[I] - \frac{\Delta t}{2}[B]} \boldsymbol{\phi}^{n} \tag{1.19}$$

for the first step and

$$\boldsymbol{\phi}^{n+1} = \frac{[I] + \frac{\Delta t}{2}[B]}{[I] - \frac{\Delta t}{2}[A]} \boldsymbol{\phi}^{n+1/2} \tag{1.20}$$

for the second step, where $\boldsymbol{\phi}^{n+1/2}$ is the intermediate field.

From (1.19), we obtain the following six equations for the first step:

$$E_x^{n+1/2} + \frac{\Delta t}{2\varepsilon}\frac{\partial H_y^{n+1/2}}{\partial z} = E_x^n + \frac{\Delta t}{2\varepsilon}\frac{\partial H_z^n}{\partial y} \tag{1.21a}$$

$$E_y^{n+1/2} + \frac{\Delta t}{2\varepsilon}\frac{\partial H_z^{n+1/2}}{\partial x} = E_y^n + \frac{\Delta t}{2\varepsilon}\frac{\partial H_x^n}{\partial z} \tag{1.21b}$$

$$E_z^{n+1/2} + \frac{\Delta t}{2\varepsilon}\frac{\partial H_x^{n+1/2}}{\partial y} = E_z^n + \frac{\Delta t}{2\varepsilon}\frac{\partial H_y^n}{\partial x} \tag{1.21c}$$

$$H_x^{n+1/2} + \frac{\Delta t}{2\mu}\frac{\partial E_z^{n+1/2}}{\partial y} = H_x^n + \frac{\Delta t}{2\mu}\frac{\partial E_y^n}{\partial z} \tag{1.21d}$$

$$H_y^{n+1/2} + \frac{\Delta t}{2\mu}\frac{\partial E_x^{n+1/2}}{\partial z} = H_y^n + \frac{\Delta t}{2\mu}\frac{\partial E_z^n}{\partial x} \tag{1.21e}$$

$$H_z^{n+1/2} + \frac{\Delta t}{2\mu}\frac{\partial E_y^{n+1/2}}{\partial x} = H_z^n + \frac{\Delta t}{2\mu}\frac{\partial E_x^n}{\partial y} \tag{1.21f}$$

Here, to eliminate $H_y^{n+1/2}$ in (1.21a), we substitute (1.21e) to (1.21a) as

$$E_x^{n+1/2} - \frac{\Delta t^2}{4\varepsilon\mu}\frac{\partial^2 E_x^{n+1/2}}{\partial z^2} = E_x^n + \frac{\Delta t}{2\varepsilon}\left(\frac{\partial H_z^n}{\partial y} - \frac{\partial H_y^n}{\partial z}\right) - \frac{\Delta t^2}{4\varepsilon\mu}\frac{\partial^2 E_z^n}{\partial x \partial z} \tag{1.22}$$

From (1.22), we obtain difference equations with three consecutive unknowns for the $E_x^{n+1/2}$ field, which can be efficiently solved using the Thomas algorithm. The other filed components are similarly solved. For the second step, the fields at $n + 1$ are obtained in the same way as in the first step. For further information regarding the ADI-FDTD method, refer to references [11–14].

1.3.3 Locally One-Dimensional Scheme

The LOD scheme has also been used to formulate the implicit FDTD method [15–20]. The difference from the ADI-FDTD method comes from the splitting scheme, that is, (1.18) is calculated as

$$\boldsymbol{\phi}^{n+1/2} = \frac{[I] + \frac{\Delta t}{2}[A]}{[I] - \frac{\Delta t}{2}[A]}\boldsymbol{\phi}^n \tag{1.23}$$

for the first step and

$$\boldsymbol{\phi}^{n+1} = \frac{[I] + \frac{\Delta t}{2}[B]}{[I] - \frac{\Delta t}{2}[B]}\boldsymbol{\phi}^{n+1/2} \tag{1.24}$$

for the second step.

From (1.23), we obtain the following six equations for the first step:

$$E_x^{n+1/2} + \frac{\Delta t}{2\varepsilon}\frac{\partial H_z^{n+1/2}}{\partial y} = E_x^n + \frac{\Delta t}{2\varepsilon}\frac{\partial H_z^n}{\partial y} \tag{1.25a}$$

$$E_y^{n+1/2} + \frac{\Delta t}{2\varepsilon}\frac{\partial H_x^{n+1/2}}{\partial z} = E_y^n + \frac{\Delta t}{2\varepsilon}\frac{\partial H_x^n}{\partial z} \tag{1.25b}$$

$$E_z^{n+1/2} + \frac{\Delta t}{2\varepsilon}\frac{\partial H_y^{n+1/2}}{\partial x} = E_z^n + \frac{\Delta t}{2\varepsilon}\frac{\partial H_y^n}{\partial x} \tag{1.25c}$$

$$H_x^{n+1/2} + \frac{\Delta t}{2\mu}\frac{\partial E_y^{n+1/2}}{\partial z} = H_x^n + \frac{\Delta t}{2\mu}\frac{\partial E_y^n}{\partial z} \tag{1.25d}$$

$$H_y^{n+1/2} + \frac{\Delta t}{2\mu}\frac{\partial E_z^{n+1/2}}{\partial x} = H_y^n + \frac{\Delta t}{2\mu}\frac{\partial E_z^n}{\partial x} \tag{1.25e}$$

$$H_z^{n+1/2} + \frac{\Delta t}{2\mu}\frac{\partial E_x^{n+1/2}}{\partial y} = H_z^n + \frac{\Delta t}{2\mu}\frac{\partial E_x^n}{\partial y} \tag{1.25f}$$

To eliminate $H_z^{n+1/2}$ in (1.25a), we substitute (1.25f) to (1.25a), leading to

$$E_x^{n+1/2} - \frac{\Delta t^2}{4\varepsilon\mu}\frac{\partial^2 E_x^{n+1/2}}{\partial y^2} = E_x^n + \frac{\Delta t}{\varepsilon}\frac{\partial H_z^n}{\partial y} - \frac{\Delta t^2}{4\varepsilon\mu}\frac{\partial^2 E_x^n}{\partial y^2} \tag{1.26}$$

As can be seen from (1.26), there exist derivatives only in the y direction. This is why it is called "locally one-dimensional" scheme. The other components for the first step and the components for the second step are similarly calculated. The advantage of the LOD-FDTD method is that the number of arithmetic operators is reduced, in comparison with the ADI counterpart [18].

For the above implementation, we have employed a two-step scheme as in the ADI scheme. Ahmed et al., have proposed the 3D LOD-FDTD method implemented in a three-step scheme [20], in which (1.17) is split into

$$\begin{aligned}
[\mathcal{A}] &= \begin{bmatrix} 0 & 0 & 0 & 0 & 0 & 0 \\ 0 & 0 & 0 & 0 & 0 & -\frac{\partial}{\varepsilon\partial x} \\ 0 & 0 & 0 & 0 & \frac{\partial}{\varepsilon\partial x} & 0 \\ 0 & 0 & 0 & 0 & 0 & 0 \\ 0 & 0 & \frac{\partial}{\mu\partial x} & 0 & 0 & 0 \\ 0 & -\frac{\partial}{\mu\partial x} & 0 & 0 & 0 & 0 \end{bmatrix} + \begin{bmatrix} 0 & 0 & 0 & 0 & 0 & \frac{\partial}{\varepsilon\partial y} \\ 0 & 0 & 0 & 0 & 0 & 0 \\ 0 & 0 & 0 & -\frac{\partial}{\varepsilon\partial y} & 0 & 0 \\ 0 & 0 & -\frac{\partial}{\mu\partial y} & 0 & 0 & 0 \\ 0 & 0 & 0 & 0 & 0 & 0 \\ \frac{\partial}{\mu\partial y} & 0 & 0 & 0 & 0 & 0 \end{bmatrix} \\
&+ \begin{bmatrix} 0 & 0 & 0 & 0 & -\frac{\partial}{\varepsilon\partial z} & 0 \\ 0 & 0 & 0 & \frac{\partial}{\varepsilon\partial z} & 0 & 0 \\ 0 & 0 & 0 & 0 & 0 & 0 \\ 0 & \frac{\partial}{\mu\partial z} & 0 & 0 & 0 & 0 \\ -\frac{\partial}{\mu\partial z} & 0 & 0 & 0 & 0 & 0 \\ 0 & 0 & 0 & 0 & 0 & 0 \end{bmatrix} \\
&= [\tilde{A}] + [\tilde{B}] + [\tilde{C}]
\end{aligned} \tag{1.27}$$

The resultant LOD calculation is obtained as

$$\boldsymbol{\phi}^{n+1/3} = \frac{[I] + \frac{\Delta t}{2}[\tilde{A}]}{[I] - \frac{\Delta t}{2}[\tilde{A}]}\boldsymbol{\phi}^n \tag{1.28}$$

for the first step,

$$\boldsymbol{\phi}^{n+2/3} = \frac{[I] + \frac{\Delta t}{2}[\tilde{B}]}{[I] - \frac{\Delta t}{2}[\tilde{B}]}\boldsymbol{\phi}^{n+1/3} \tag{1.29}$$

for the second step and

$$\boldsymbol{\phi}^{n+1} = \frac{[I] + \frac{\Delta t}{2}[\tilde{C}]}{[I] - \frac{\Delta t}{2}[\tilde{C}]}\boldsymbol{\phi}^{n+2/3} \tag{1.30}$$

for the third step. The finite-difference equations are derived similarly.

Note that the original LOD-FDTD method is first-order accurate in time. Tan has proposed the 3D LOD-FDTD method with a second-order temporal accuracy [21].

1.3.4 Fundamental Scheme

The fundamental scheme [22] is used to efficiently implement the ADI- and LOD-FDTD methods, in which a matrix-free formulation can be done in the right-hand side of the resultant equations. The important point is that the simple formulation by the fundamental scheme is completely equivalent to that of the conventional implicit formulation. For instance, given us the application of the fundamental scheme to the LOD-FDTD method.

The LOD-FDTD equations are as follows:

$$\left([I] - \frac{\Delta t}{2}[A]\right)\boldsymbol{\phi}^{n+1/2} = \left([I] + \frac{\Delta t}{2}[A]\right)\boldsymbol{\phi}^{n} \tag{1.31a}$$

$$\left([I] - \frac{\Delta t}{2}[B]\right)\boldsymbol{\phi}^{n+1} = \left([I] + \frac{\Delta t}{2}[B]\right)\boldsymbol{\phi}^{n+1/2} \tag{1.31b}$$

Equation (1.31a) can be rewritten as

$$\begin{aligned}\left([I] - \frac{\Delta t}{2}[A]\right)\boldsymbol{\phi}^{n+1/2} &= \left([I] + \frac{\Delta t}{2}[A]\right)\boldsymbol{\phi}^{n} \\ &= 2\boldsymbol{\phi}^{n} - \left([I] - \frac{\Delta t}{2}[A]\right)\boldsymbol{\phi}^{n+1/2}\end{aligned} \tag{1.32}$$

This can be manipulated readily to give

$$\left([I] - \frac{\Delta t}{2}[A]\right)\left(\boldsymbol{\phi}^{n+1/2} + \boldsymbol{\phi}^{n}\right) = 2\boldsymbol{\phi}^{n}$$

where the vector terms in the bracket can be denoted by an auxiliary variable

$$\boldsymbol{v}^{n+1/2} = \boldsymbol{\phi}^{n+1/2} + \boldsymbol{\phi}^{n}$$

Similar manipulation can be applied to (1.31b). As a result, we derive the following simplified equations for the LOD-FDTD method:

$$\left([I] - \frac{\Delta t}{2}[A]\right)\boldsymbol{v}^{n+1/2} = \boldsymbol{\phi}^{n} \tag{1.33}$$

$$\boldsymbol{\phi}^{n+1/2} = \boldsymbol{v}^{n+1/2} - \boldsymbol{\phi}^{n} \tag{1.34}$$

for the first step and

$$\left([I] - \frac{\Delta t}{2}[B]\right)\boldsymbol{v}^{n+1} = \boldsymbol{\phi}^{n+1/2} \tag{1.35}$$

$$\boldsymbol{\phi}^{n+1} = \boldsymbol{v}^{n+1} - \boldsymbol{\phi}^{n+1/2} \tag{1.36}$$

for the second step. As can be seen the above fundamental equations do not have spatial derivatives in the right-hand sides, which leads to an efficient implementation of the LOD-FDTD method.

1.4 Hybrid-Implicit-Explicit Technique

The implicit FDTD methods discussed above have the advantage that time step size can be chosen arbitrarily large even when the spatial sampling widths are quite small in all three directions. This contributes significantly to reducing the computation time. Note, however, that the implicit FDTD methods are not always efficient for the case where the sampling width is chosen small only in a specific direction. In this case, the HIE-FDTD method becomes efficient [23–29], in which the implicit scheme is applied only to this specific direction of small sampling widths.

For example, when the small sampling width is required only for the y direction, the field in the y direction is averaged in time. Then, (1.3)–(1.8) are rewritten as

$$\mu\frac{\partial H_x^{n+1/2}}{\partial t} = \frac{\partial E_y^{n+1/2}}{\partial z} - \frac{\partial}{\partial y}\frac{E_z^{n} + E_z^{n+1}}{2} \tag{1.37}$$

$$\mu\frac{\partial H_y^{n}}{\partial t} = \frac{\partial E_z^{n}}{\partial x} - \frac{\partial E_x^{n}}{\partial z} \tag{1.38}$$

$$\mu\frac{\partial H_z^{n+1/2}}{\partial t} = \frac{\partial}{\partial y}\frac{E_x^{n} + E_x^{n+1}}{2} - \frac{\partial E_y^{n+1/2}}{\partial x} \tag{1.39}$$

$$\varepsilon\frac{\partial E_x^{n+1/2}}{\partial t} = \frac{\partial H_z^{n+1/2}}{\partial y} - \frac{\partial H_y^{n+1/2}}{\partial z} \tag{1.40}$$

$$\varepsilon \frac{\partial E_y^n}{\partial t} = \frac{\partial H_x^n}{\partial z} - \frac{\partial H_z^n}{\partial x} \tag{1.41}$$

$$\varepsilon \frac{\partial E_z^{n+1/2}}{\partial t} = \frac{\partial H_y^{n+1/2}}{\partial x} - \frac{\partial}{\partial y}\frac{H_x^{n+1} + H_x^n}{2} \tag{1.42}$$

in which the central position of time for the discretization is chosen as follows: $H_x^{n+1/2}$, H_y^n, $H_z^{n+1/2}$, $E_x^{n+1/2}$, E_y^n and $E_z^{n+1/2}$. The term regarding time derivatives is approximated as

$$H_x^{n+1} = H_x^n + \frac{\Delta t}{\mu}\left(\frac{\partial E_y^{n+1/2}}{\partial z} - \frac{\partial}{\partial y}\frac{E_z^n + E_z^{n+1}}{2}\right) \tag{1.43}$$

$$H_y^{n+1/2} = H_y^{n-1/2} + \frac{\Delta t}{\mu}\left(\frac{\partial E_z^n}{\partial x} - \frac{\partial E_x^n}{\partial z}\right) \tag{1.44}$$

$$H_z^{n+1} = H_z^n + \frac{\Delta t}{\mu}\left(\frac{\partial}{\partial y}\frac{E_x^n + E_x^{n+1}}{2} - \frac{\partial E_y^{n+1/2}}{\partial x}\right) \tag{1.45}$$

$$E_x^{n+1} = E_x^n + \frac{\Delta t}{\varepsilon}\left(\frac{\partial}{\partial y}\frac{H_z^{n+1} + H_z^n}{2} - \frac{\partial H_y^{n+1/2}}{\partial z}\right) \tag{1.46}$$

$$E_y^{n+1/2} = E_y^{n-1/2} + \frac{\Delta t}{\varepsilon}\left(\frac{\partial H_x^n}{\partial z} - \frac{\partial H_z^n}{\partial x}\right) \tag{1.47}$$

$$E_z^{n+1} = E_z^n + \frac{\Delta t}{\varepsilon}\left(\frac{\partial H_y^{n+1/2}}{\partial x} - \frac{\partial}{\partial y}\frac{H_x^{n+1} + H_x^n}{2}\right) \tag{1.48}$$

First, we calculate (1.44) and (1.47) explicitly. Then, (1.43) and (1.45) are respectively substituted to (1.48) and (1.46), and they are solved implicitly. Finally, (1.43) and (1.45) are solved explicitly. The CFL condition of the above HIE-FDTD method is expressed as

$$v\Delta t \leq \frac{1}{\sqrt{\left(\frac{1}{\Delta x}\right)^2 + \left(\frac{1}{\Delta z}\right)^2}} \tag{1.49}$$

Note that the condition is determined with the sampling widths in the x and z directions, leading to a weakly conditionally stable FDTD method. The HIE technique is also applied to the FDTD method in cylindrical coordinates.

1.5 Convolutional Perfectly Matched Layer

The conventional PML requires the split-field formulation which is different from the original single-field formulation of the FDTD method. A stretch-coordinate formulation can be performed without using the split-field one. Here, we introduce the convolutional PML (CPML) with the help of the stretch-coordinate formulation [30].

For simplicity, we treat the one-dimensional problem expressed as follows:

$$\varepsilon \frac{\partial E_y}{\partial t} = \frac{\partial H_x}{\partial z} \tag{1.50}$$

Equation (1.50) is transformed into the one in the frequency domain as

$$j\omega\varepsilon E_y = \frac{\partial H_x}{\partial z} \tag{1.51}$$

We introduce a lossy coefficient into (1.51) for the absorbing region as

$$j\omega\varepsilon E_y = \frac{1}{s_z}\frac{\partial H_x}{\partial z} \tag{1.52}$$

where

$$s_z = k_z + \frac{\sigma_z}{\alpha_z + j\omega\varepsilon_0} \tag{1.53}$$

in which α_z and σ_z are assumed to be positive and k_z is real and larger than 1. Note that α_z and k_z contribute to absorbing evanescent waves, while $\alpha_z = 0$ and $k_z = 1$ for the original PML.

Now, we transform (1.52) to the one of the convolution form in the time domain

$$\varepsilon \frac{\partial E_y}{\partial t} = \bar{s}_z * \frac{\partial H_x}{\partial z} \tag{1.54}$$

where $\bar{s}_z$ is the Laplace transform of s_z^{-1} as

$$\begin{aligned} \bar{s}_z(t) &= \frac{\delta(t)}{k_z} - \frac{\sigma_z}{\varepsilon_0 k_z^2} e^{((\sigma_z/\varepsilon_0 k_z)+(\alpha_z/\varepsilon_0))t} u(t) \\ &= \frac{\delta(t)}{k_z} + \zeta_z(t) \end{aligned} \tag{1.55}$$

Substituting (1.55) into (1.54) results in

$$\varepsilon \frac{\partial E_y}{\partial t} = \frac{1}{k_z}\frac{\partial H_x}{\partial z} + \zeta_z(t) * \frac{\partial H_x}{\partial z} \tag{1.56}$$

The convolution calculation is performed as follows:

$$
\begin{aligned}
Z_{0z}(m) &= \int_{m\Delta t}^{(m+1)\Delta t} \zeta_z(\tau)d\tau \\
&= a_z e^{((\sigma_z/k_z)+\alpha_z)(m\Delta t/\varepsilon_0)}
\end{aligned} \tag{1.57}
$$

where

$$
a_z = \frac{\sigma_z}{\sigma_z k_z + k_z^2 \alpha_z}(e^{-((\sigma_z/k_z)+\alpha_z)(\Delta t/\varepsilon_0)} - 1) \tag{1.58}
$$

As a result, (1.56) can be expressed as

$$
\begin{aligned}
\varepsilon\frac{E_{y,k}^{n+1} - E_{y,k}^{n}}{\Delta t} &= \frac{H_{x,k+1/2}^{n+1/2} - H_{x,k-1/2}^{n+1/2}}{\Delta z} + \sum_{m=0}^{N-1} Z_{0z}(m)\frac{H_{x,k+1/2}^{n-m+1/2} - H_{x,k-1/2}^{n-m+1/2}}{\Delta z} \\
&= \frac{H_{x,k+1/2}^{n+1/2} - H_{x,k-1/2}^{n+1/2}}{\Delta z} + \Phi^{n+1/2}
\end{aligned} \tag{1.59}
$$

Using (1.57) and (1.58), the application of the recursive convolution (RC) technique leads to

$$
\Phi^{n+1/2} = b_z \Phi^{n-1/2} + a_z \frac{H_{x,k+1/2}^{n+1/2} - H_{x,k-1/2}^{n+1/2}}{\Delta z} \tag{1.60}
$$

where

$$
b_z = e^{-((\sigma_z/k_z)+\alpha_z)(\Delta t/\varepsilon_0)} \tag{1.61}
$$

The similar formulation can be applied to the absorbing regions in the x and y directions.

1.6 Concluding Remarks

The formulation of the FDTD method has been presented and its application to the implicit FDTD method has also been provided. To implement the implicit FDTD method efficiently, the fundamental scheme has been applied to the LOD formulation. Furthermore, the HIE-FDTD method has been explained for the problem where small meshes are required only in the specific direction. As an example of an absorbing material, the formulation of the CPML has been given for a one-dimensional problem.

References

1. K.S. Yee, Numerical solution of initial boundary value problems involving Maxwell's equations in isotropic media. IEEE Trans. Antennas Propagat. **AP-14**, 302–307 (1966)
2. A. Taflove, S.C. Hagness, *Computational Electrodynamics: The Finite-Difference Time-Domain Method*, 3rd edn. (Artech House, Norwood, MA, USA, 2005)
3. A. Taflove, A. Oskooi, S.G. Johnson, *Advances in FDTD Computational Electrodynamics: Photonics and Nanotechnology* (Artech House, Boston, Mass, 2013)
4. Y. Yang, R.S. Chen, E.K.N. Yung, The unconditionally stable Crank Nicolson FDTD method for three-dimensional Maxwell's equations. Microw. Opt. Tech. Lett. **48**(8), 1619–1622 (2006)
5. G. Sun, C.W. Trueman, Unconditionally-stable FDTD method based on Crank-Nicolson scheme for solving three-dimensional Maxwell equations. Electron. Lett. **40**(10), 589–590 (2004)
6. E.L. Tan, Efficient algorithms for Crank-Nicolson-based finite-difference time-domain methods. IEEE Trans. Microw. Theory Techn. **56**(2), 408–413 (2008). https://doi.org/10.1109/TMTT.2007.914641
7. T. Namiki, A new FDTD algorithm based on alternating-direction implicit method. IEEE Trans. Microw. Theory Tech. **47**(10), 2003–2007 (1999)
8. F.H. Zheng, Z.Z. Chen, J.Z. Zhang, A finite-difference time-domain method without the Courant stability conditions. IEEE Microw. Guided Wave Lett. **9**(11), 441–443 (1999)
9. F. Zhen, Z.Z. Chen, J.Z. Zhang, Toward the development of a three-dimensional unconditionally stable finite-difference time-domain method. IEEE Trans. Microw. Theory Tech. **48**(9), 1550–1558 (2000)
10. T. Namiki, 3-D ADI-FDTD method-unconditionally stable time-domain algorithm for solving full vector Maxwell's equations. IEEE Trans. Microw. Theory Tech. **48**(10), 1743–1748 (2000)
11. S.G. Garcia, T.W. Lee, S.C. Hagness, On the accuracy of the ADI-FDTD method. IEEE Antennas Wireless Propagat. Lett. **1**, 31–34 (2002)
12. A.P. Zhao, Analysis of the numerical dispersion of the 2D alternating-direction implicit FDTD method. IEEE Trans. Microw. Theory Tech. **50**(4), 1156–1164 (2002)
13. H. Rao, R. Scarmozzino, R.M. Osgood Jr., An improved ADI-FDTD method and its application to photonic simulations. IEEE Photon. Technol. Lett. **14**(4), 477–479 (2002)
14. S.G. Garcia, R.G. Rubio, A.R. Bretones, R.G. Lopez, Revisiting the stability of Crank-Nicolson and ADI-FDTD. IEEE Trans. Antennas Propag. **55**(11), 3199–3203 (2007)
15. J. Shibayama, M. Muraki, J. Yamauchi, H. Nakano, Efficient implicit FDTD algorithm based on locally one-dimensional scheme. Electron. Lett. **41**(19), 1046–1047 (2005)
16. J. Shibayama, M. Muraki, R. Takahashi, J. Yamauchi, H. Nakano, Performance evaluation of several implicit FDTD methods for optical waveguide analyses. J. Lightw. Technol. **24**(6), 2465–2472 (2006)
17. V.E. Nascimento, B.-H.V. Borges, F.L. Teixeira, Split-field PML implementations for the unconditionally stable LOD-FDTD method. IEEE Microw. Wireless Compon. Lett. **16**(7), 398–400 (2006)
18. E.L. Tan, Unconditionally stable LOD-FDTD method for 3-D Maxwell's equations. IEEE Microw. Wireless Compon. Lett. **17**(2), 85–87 (2007)
19. E. Li, I. Ahmed, R. Vahldieck, Numerical dispersion analysis with an improved LOD-FDTD method. IEEE Microw. Wireless Compon. Lett. **17**(5), 319–321 (2007)
20. I. Ahmed, E.K. Chua, E.P. Li, Z. Chen, Development of the three-dimensional unconditionally stable LOD-FDTD method. IEEE Trans. Antennas Propagat. **56**(11), 3596–3600 (2008)
21. T.H. Gan, E.L. Tan, Unconditionally stable fundamental LOD-FDTD method with second-order temporal accuracy and complying divergence. IEEE Trans. Antennas Propagat. **61**(5), 2630–2638 (2013)
22. E.L. Tan, Fundamental schemes for efficient unconditionally stable implicit finite-difference time-domain methods. IEEE Trans. Antennas Propagat. **56**(1), 170–177 (2008)
23. B. Huang, G. Wang, Y. Jiang, W. Wang, A hybrid implicit-explicit FDTD scheme with weakly conditional stability. Microw. Opt. Tech. Lett. **39**(2), 97–101 (2003)

24. J. Chen, J. Wang, A three-dimensional semi-implicit FDTD scheme for calculation of shielding effectiveness of enclosure with thin slots. IEEE Trans. Electromagn. Compat. **49**(2), 354–360 (2007)
25. M. Unno, H. Asai, HIE-FDTD method for hybrid system with lumped elements and conductive media. IEEE Microw. Wireless Compon. Lett. **21**(9), 453–455 (2011)
26. J. Wang, B. Zhou, L. Shi, C. Gao, B. Chen, A novel 3-D HIE-FDTD method with one-step leapfrog scheme. IEEE Trans. Microw. Theory Techn. **62**(6), 1275–1283 (2014)
27. K. Fujita, Hybrid Newmark-conformal FDTD method for multiphysics modeling of short spark gaps with curved metallic surfaces. IEEE J. Multiscale Multiphys. Comput. Techn. **2**, 66–77 (2017)
28. A. Van Londersele, D. De Zutter, D. Vande Ginste, A collocated 3-D HIE-FDTD scheme with PML. IEEE Microw. Wireless Compon. Lett. **27**(7), 609–611 (2017)
29. J. Chen, J. Wang, C. Tian, Three-dimensional hybrid implicit-explicit finite-difference time-domain method in the cylindrical coordinate system. IET Microw. Antennas Propag. **3**(8), 1254–1261 (2009)
30. J.A. Roden, S.D. Gedney, Convolution PML (CPML): an efficient FDTD implementation of the CFS-PML for arbitrary media. Microw. Opt. Tech. Lett. **27**(5), 334–339 (2000)

Chapter 2
Frequency-Dependent FDTD Method

Abstract This chapter presents the details of the frequency-dependent formulations for the analysis of plasmonic devices consisting of dispersive media. Three frequently-used techniques are given such as the auxiliary differential equation (ADE), recursive convolution (RC)-related and Z-transform (ZT) techniques. The ADE technique is based on the use of differential equation explaining the relation between D and E. The D-E equation in time domain is discretized by the finite-difference scheme and simultaneously solved with the original FDTD equations. In the RC-related techniques, the convolution calculation between the permittivity and the electric field can be efficiently treated with a recursive procedure. The original RC technique is first-order accurate with one convolution calculation, while the piecewise linear RC (PLRC) technique is second-order accurate with two convolution calculations. The trapezoidal RC (TRC) technique is second-order accurate even with one convolution calculation. In the ZT technique, the complicated convolution integrals can simply be reduced to algebraic equations, and the D-E relation can be transformed into finite-difference equations. Here, these three techniques are explained for the Drude and Drude-Lorentz models. In addition, the Drude-Critical points (D-CP) model is also investigated with the TRC technique. The application to the implicit LOD-FDTD method is also discussed and the accuracy of the frequency-dependent formulations is compared.

2.1 Introduction

Since the FDTD method is a time-domain technique, it is relatively cumbersome to handle frequency-dependent or frequency-dispersive media [1–3] often applied to plasmonic devices. There are three types of frequency-dependent FDTD formulations, such as the auxiliary differential equation (ADE) [4, 5], Z-transformation (ZT) [6], and recursive convolution (RC)-related [7–10] techniques. In this section, we introduce these three frequency-dependent formulations into the FDTD method. In particular, the formulations of the commonly used dispersion models are given for the Drude, Drude-Lorentz, and Drude-Critical points (D-CP) models. A comparative

J. Shibayama, *FDTD Analysis of Plasmonic Devices*, Springer Series in Optical Sciences 254, https://doi.org/10.1007/978-981-96-8130-3_2

study is performed for these three dispersion models, in which the implicit FDTD method based on the locally one-dimensional (LOD) scheme is also evaluated in comparison to the traditional explicit FDTD method.

2.2 Dispersion Model

The Drude–Lorentz model [1] (with P poles for the Lorentz function) has widely been used to account for the measured permittivities of Au and Ag over a wide spectrum from the visible to near infrared regions as follows:

$$\varepsilon_r(\omega) = \varepsilon_\infty + \frac{\omega_{\mathrm{D}}^2}{j\omega\nu_{\mathrm{D}} - \omega^2} + \sum_{p=1}^{P} \frac{\Delta\varepsilon G_p \omega_p^2}{j\omega\nu_p + \omega_p^2 - \omega^2} \tag{2.1}$$

where ε_∞ is the dielectric constant of the material at infinite frequency, ω is the angular frequency, ω_{D} and ω_p are the electron plasma frequencies, ν_{D} and ν_p are the effective electron collision frequencies, G_p is the pole strength, and $\Delta\varepsilon$ is the weighting coefficient. Eliminating the last term in the right-hand side of (2.1) results in the Drude model.

The D-CP model are often used to express dispersion characteristics of a specific metal, expressed as [2, 3]

$$\varepsilon_r(\omega) = \varepsilon_\infty + \frac{\omega_{\mathrm{D}}^2}{j\omega\nu_{\mathrm{D}} - \omega^2} + \sum_{p=1}^{P} \bar{\chi}_p(\omega) \tag{2.2}$$

where

$$\bar{\chi}_p(\omega) = \Delta\varepsilon_p \omega_p \left(\frac{e^{-j\rho_p}}{\omega_p - \omega + j\nu_p} + \frac{e^{j\rho_p}}{\omega_p + \omega - j\nu_p} \right) \tag{2.3}$$

in which ρ_p is the phase, and $\Delta\varepsilon$ is the weighting coefficient. Equation (2.2) gives a better fitting model for the function possessing critical points with less additional terms P, when compared with the Drude–Lorentz model.

2.3 ADE (Auxiliary Differential Equation) Technique

Here, the formulation of the ADE technique is presented for the Drude–Lorentz model, as an example. The ADE technique can avoid the calculations of the convolution integrals used in the RC-related techniques shown below. In the ADE technique [4, 5], the D-E relation in frequency domain is translated into that in time domain using the inverse Fourier transform.

The E-H relation in frequency domain is expressed as

$$\frac{j\omega\varepsilon_r(\omega)}{c}\boldsymbol{E} = \nabla \times \boldsymbol{H} \tag{2.4}$$

where c is the speed of light in a vacuum. After substituting (2.1) into (2.24) and applying the inverse Fourier transform to the resultant equation ($j\omega \to \partial/\partial t$), we obtain

$$\varepsilon_\infty \frac{\partial \boldsymbol{E}}{\partial t} + \boldsymbol{J} + \sum_{p=1}^{P} \frac{\partial \boldsymbol{Q}_p}{\partial t} = c\nabla \times \boldsymbol{H} \tag{2.5}$$

$$\frac{\partial \boldsymbol{J}}{\partial t} + \nu_{\mathrm{D}} \boldsymbol{J} = \omega_{\mathrm{D}}^2 \boldsymbol{E} \tag{2.6}$$

$$\frac{\partial^2 \boldsymbol{Q}_p}{\partial t^2} + \nu_p \frac{\partial \boldsymbol{Q}_p}{\partial t} + \omega_p^2 \boldsymbol{Q}_p = \Delta\varepsilon G_p \omega_p^2 \boldsymbol{E}. \tag{2.7}$$

These equations are discretized with the finite-difference scheme and the resultant equations are solved simultaneously with the standard FDTD equations.

2.4 Z-Transform Technique

Another way of avoiding the calculations of the convolution integrals is to use the ZTs [6]. In this technique, the complicated convolution integrals, treated in the RC-based techniques below, can simply be reduced to algebraic equations, and the D-E relation can be translated into finite-difference equations in a straightforward manner, compared with the ADE technique.

To apply the ZT theory, we rewrite (2.1) as

$$\varepsilon_r(\omega) = \varepsilon_\infty + \frac{\omega_{\mathrm{D}}^2/\nu_{\mathrm{D}}}{j\omega} - \frac{\omega_{\mathrm{D}}^2/\nu_{\mathrm{D}}}{\nu_{\mathrm{D}} + j\omega} + \sum_{p=1}^{P} \frac{\Delta\varepsilon G_p \omega_p^2 \beta/\beta}{\alpha^2 + \beta^2 + j2\alpha\omega + (j\omega)^2}. \tag{2.8}$$

Taking the Z transform of (2.8) and $\boldsymbol{D}(\omega) = \varepsilon_r(\omega)\boldsymbol{E}(\omega)$ (with the field normalization) yields the following equations in the Z domain:

$$\boldsymbol{D}(z) = \varepsilon_\infty \boldsymbol{E}(z) + z^{-1}\boldsymbol{M}(z) + z^{-1}\sum_{p=1}^{P} \boldsymbol{Q}_p(z)$$

$$\boldsymbol{M}(z) = \frac{\frac{\omega_{\mathrm{D}}^2 \Delta t}{\nu_{\mathrm{D}}}\left(1 - e^{-\nu_{\mathrm{D}}\Delta t}\right)}{1 - \left(1 + e^{-\nu_{\mathrm{D}}\Delta t}\right) z^{-1} + e^{-\nu_{\mathrm{D}}\Delta t} z^{-2}} \boldsymbol{E}(z)$$

$$Q_p(z) = \frac{\frac{\Delta\varepsilon G_p \omega_p^2 \Delta t}{\beta} e^{-\alpha\Delta t} \sin(\beta\Delta t)}{1 - 2e^{-\alpha\Delta t}\cos(\beta\Delta t) z^{-1} + e^{-2\alpha\Delta t} z^{-2}} \boldsymbol{E}(z)$$

in which Δt and z^{-l} represent the time period and the delay of l time periods, respectively. We readily translate the above into the finite-difference equations in the Z domain as

$$\boldsymbol{E}^{n+1} = \frac{1}{\varepsilon_\infty}\boldsymbol{D}^{n+1} - \frac{1}{\varepsilon_\infty}\boldsymbol{M}^n - \frac{1}{\varepsilon_\infty}\sum_{p=1}^{P} \boldsymbol{Q}_p^n \tag{2.9}$$

$$\boldsymbol{M}^{n+1} = (1 + e^{-\nu_{\mathrm{D}}\Delta t})\boldsymbol{M}^n - e^{-\nu_{\mathrm{D}}\Delta t}\boldsymbol{M}^{n-1} + \frac{\omega_{\mathrm{D}}^2 \Delta t}{\nu_{\mathrm{D}}}(1 - e^{-\nu_{\mathrm{D}}\Delta t})\boldsymbol{E}^{n+1} \tag{2.10}$$

$$\boldsymbol{Q}_p^{n+1} = 2e^{-\alpha\Delta t}\cos(\beta\Delta t)\,\boldsymbol{Q}_p^n - e^{-2\alpha\Delta t}\,\boldsymbol{Q}_p^{n-1} + \frac{\Delta\varepsilon G_p \omega_p^2 \Delta t}{\beta} e^{-\alpha\Delta t}\sin(\beta\Delta t)\boldsymbol{E}^{n+1}. \tag{2.11}$$

Note that the above equations regarding the D-E relation are analytically derived without adopting the finite-difference approximation used in the ADE technique. We solve (2.9)–(2.11), combined with the standard FDTD equations.

2.5 RC (Recursive Convolution)-Related Techniques

The RC-based techniques are widely used to formulate the frequency-dependent FDTD method [7–10], since the calculation of the convolution integral can be efficiently performed. Note that the techniques are applicable only to linear dispersive materials.

The D-E relation with (2.1) is written in the frequency-domain as

$$\boldsymbol{D}(\omega) = \varepsilon_0 \varepsilon_r(\omega)\boldsymbol{E}(\omega) \tag{2.12}$$

where ε_0 is the permittivity of free space. Taking the inverse Fourier transform of (2.12), we obtain the following convolution form:

$$\boldsymbol{D}(t) = \int_{-\infty}^{\infty} \varepsilon_0 \varepsilon_r(\tau)\boldsymbol{E}(t-\tau)d\tau \tag{2.13}$$

As a result, the D-E relation in time domain is expressed for the discretization form as

$$\begin{aligned} \boldsymbol{D}(t) &= \varepsilon_0\varepsilon_\infty \boldsymbol{E}(t) + \varepsilon_0 \int_0^{n\Delta t} \boldsymbol{E}(n\Delta t - \tau)\chi(\tau)d\tau \\ &\equiv \varepsilon_0\varepsilon_\infty \boldsymbol{E}(t) + \varepsilon_0 \boldsymbol{P}(n\Delta t) \end{aligned} \tag{2.14}$$

where $\boldsymbol{P}$ is the linear polarization and $\chi(\tau)$ is susceptibility. The RC-related techniques are methods for efficiently handling this convolutional computation on $\boldsymbol{P}$.

For the RC technique [7], $\boldsymbol{P}$ is approximated as

$$\boldsymbol{P}(n\Delta t) = \sum_{m=0}^{n-1} \boldsymbol{E}^{n-m}\chi^{m} \tag{2.15}$$

where

$$\chi^{m} = \int_{m\Delta t}^{(m+1)\Delta t} \chi(\tau)d\tau \tag{2.16}$$

in which the electric field is assumed constant over Δt.

For the PLRC technique [8], $\boldsymbol{P}$ is approximated as

$$\boldsymbol{P}(n\Delta t) = \sum_{m=0}^{n-1} \{\boldsymbol{E}^{n-m}\chi^{m} + (\boldsymbol{E}^{n-m-1} - \boldsymbol{E}^{n-m})\xi^{m}\} \tag{2.17}$$

where

$$\xi^{m} = \frac{1}{\Delta t}\int_{m\Delta t}^{(m+1)\Delta t} (\tau - m\Delta\tau)\chi(\tau)d\tau$$

in which the electric field has piecewise linear functional dependence over Δt. Note that the term regarding ξ^m appears compared with (2.15), requiring an additional calculation of the convolution integral. This formulation results in storage of electric field values from two time steps [8].

With the PLRC technique, we obtain the following equation of the electric field (the normalized expression of field components is used):

$$\boldsymbol{E}^{n+1} = \frac{\varepsilon_\infty - \xi^0}{\varepsilon_\infty + \chi^0 - \xi^0}\boldsymbol{E}^{n} + \frac{1}{\varepsilon_\infty + \chi^0 - \xi^0}\boldsymbol{\phi}^{n} + \frac{c\Delta t}{\varepsilon_\infty + \chi^0 - \xi^0}\nabla \times \boldsymbol{H}^{n+1/2} \tag{2.18}$$

where $\chi^0 = \chi_{\mathrm{D}}^0 + \sum_{p=1}^{P} \mathrm{Re}\left[\chi_p^0\right]$, and $\boldsymbol{\phi}^n = \boldsymbol{\phi}_{\mathrm{D}}^n + \sum_{p=1}^{P} \mathrm{Re}\left[\boldsymbol{\phi}_p^n\right]$. $\boldsymbol{\phi}_{\mathrm{D}}^n$ and $\boldsymbol{\phi}_p^n$ are expressed as follows:

$$\boldsymbol{\phi}_{\mathrm{D}}^{n} = \left(\Delta\chi_{\mathrm{D}}^{0} - \Delta\xi_{\mathrm{D}}^{0}\right)\boldsymbol{E}^{n} + \Delta\xi_{\mathrm{D}}^{0}\boldsymbol{E}^{n-1} + e^{-\nu_{\mathrm{D}}\Delta t}\boldsymbol{\phi}_{\mathrm{D}}^{n-1}$$
$$\boldsymbol{\phi}_{p}^{n} = \left(\Delta\chi_{p}^{0} - \Delta\xi_{p}^{0}\right)\boldsymbol{E}^{n} + \Delta\xi_{p}^{0}\boldsymbol{E}^{n-1} + e^{\gamma\Delta t}\boldsymbol{\phi}_{p}^{n-1}$$

where $\gamma = -\alpha + j\beta$, $\alpha = \nu_p/2$, and $\beta = \sqrt{\omega_p^2 - \alpha^2}$. The parameters used in the above formulation are given as

$$
\begin{aligned}
\chi_{\mathrm{D}}^{0} &= \frac{\omega_{\mathrm{D}}^{2}}{\nu_{\mathrm{D}}}\left\{\Delta t - \frac{1}{\nu_{\mathrm{D}}}(1 - e^{-\nu_{\mathrm{D}}\Delta t})\right\} \\
\chi_{p}^{0} &= -\frac{2\Delta\varepsilon G_{p}\omega_{p}^{2}}{\gamma(\gamma - \gamma^{*})}(1 - e^{\gamma\Delta t}) \\
\Delta\chi_{\mathrm{D}}^{0} &= -\frac{\omega_{\mathrm{D}}^{2}}{\nu_{\mathrm{D}}^{2}}(1 - e^{-\nu_{\mathrm{D}}\Delta t})^{2} \\
\Delta\chi_{p}^{0} &= \chi_{p}^{0}(1 - e^{\gamma\Delta t}) \\
\xi_{\mathrm{D}}^{0} &= \frac{\omega_{\mathrm{D}}^{2}}{\nu_{\mathrm{D}}}\left\{\frac{\Delta t}{2} - \frac{1}{\nu_{\mathrm{D}}^{2}\Delta t}\left(1 - e^{-\nu_{\mathrm{D}}\Delta t}\right) + \frac{1}{\nu_{\mathrm{D}}}e^{-\nu_{\mathrm{D}}\Delta t}\right\} \\
\xi_{p}^{0} &= -j\frac{\Delta\varepsilon G_{p}\omega_{p}^{2}}{\gamma^{2}\beta\Delta t}\left\{(\gamma\Delta t - 1)\, e^{\gamma\Delta t} + 1\right\} \\
\Delta\xi_{\mathrm{D}}^{0} &= -\frac{\omega_{\mathrm{D}}^{2}}{\nu_{\mathrm{D}}^{2}}\left\{\frac{1}{\nu_{\mathrm{D}}\Delta t}\left(1 - e^{-\nu_{\mathrm{D}}\Delta t}\right)^{2}\right. \\
&\qquad \left. - \left(1 - e^{-\nu_{\mathrm{D}}\Delta t}\right) e^{-\nu_{\mathrm{D}}\Delta t}\right\} \\
\Delta\xi_{p}^{0} &= \xi_{p}^{0}\left(1 - e^{\gamma\Delta t}\right).
\end{aligned}
$$

Comparison between (2.15) and (2.17) shows that neglecting the terms regarding ξ in (2.18) results in the equations for the RC technique.

As an alternative, the TRC technique was developed for the frequency-dependent formulation [9, 10]. The advantage of the TRC technique lies in the fact that it requires only a single convolution integral maintaining the accuracy comparable to the PLRC counterpart with two convolution integrals.

For the TRC technique, $\boldsymbol{P}$ is approximated using an average of the electric fields over two consecutive time steps in the following form:

$$
\boldsymbol{P}(n\Delta t) = \sum_{m=0}^{n-1} \frac{\boldsymbol{E}^{n-m} + \boldsymbol{E}^{n-m-1}}{2}\chi^{m}. \tag{2.19}
$$

Obviously, the TRC technique requires only a single convolution integral regarding χ^{m}, as in the RC technique. As a result of the formulation similar to the RC procedure, we obtain the TRC-FDTD equations as

$$
\boldsymbol{E}^{n+1} = \frac{\varepsilon_{\infty} - \chi^{0}/2}{\varepsilon_{\infty} + \chi^{0}/2}\boldsymbol{E}^{n} + \frac{1}{\varepsilon_{\infty} + \chi^{0}/2}\boldsymbol{\phi}^{n} + \frac{c\Delta t}{\varepsilon_{\infty} + \chi^{0}/2}\nabla \times \boldsymbol{H}^{n+1/2} \tag{2.20}
$$

where $\boldsymbol{\phi}^{n} = \boldsymbol{\phi}_{\mathrm{D}}^{n} + \sum_{p=1}^{P} \mathrm{Re}\left[\boldsymbol{\phi}_{p}^{n}\right]$. $\boldsymbol{\phi}_{\mathrm{D}}^{n}$ and $\boldsymbol{\phi}_{p}^{n}$ are expressed as follows:

$$
\boldsymbol{\phi}_{\mathrm{D}}^{n} = \frac{\boldsymbol{E}^{n} + \boldsymbol{E}^{n-1}}{2}\Delta\chi_{\mathrm{D}}^{0} + e^{-\nu_{\mathrm{D}}\Delta t}\boldsymbol{\phi}_{\mathrm{D}}^{n-1} \tag{2.21}
$$

$$\boldsymbol{\phi}_p^n = \frac{\boldsymbol{E}^n + \boldsymbol{E}^{n-1}}{2}\Delta\chi_p^0 + e^{\gamma\Delta t}\boldsymbol{\phi}_p^{n-1}. \tag{2.22}$$

Clearly, the equation for the TRC technique is simpler than that for the PLRC technique.

Final consideration of the TRC formulation is given for the D-CP model. To apply the Laplace transform, we rewrite (2.2) with $s = j\omega$ as

$$\bar{\chi}_p(\omega) = 2\Delta\varepsilon_p\omega_p\left[\frac{\omega_p\cos(\rho_p)}{\omega_p^2 + \left(s + \nu_p\right)^2} - \frac{(s + \nu_p)\sin(\rho_p)}{\omega_p^2 + \left(s + \nu_p\right)^2}\right]. \tag{2.23}$$

The Laplace transform of (2.23) results in

$$\bar{\chi}_p(t) = 2\Delta\varepsilon_p\omega_p e^{-\nu_p t}\sin(\omega_p t - \rho_p). \tag{2.24}$$

Unfortunately, the term regarding $\sin(\omega_p t - \rho_p)$ prevents us from directly deriving a recursive expression. To obtain the recursive expression, we introduce the following complex susceptibility:

$$\tilde{\chi}_p(t) = 2j\Delta\varepsilon_p\omega_p e^{j\rho_p}e^{-\gamma t} \tag{2.25}$$

where $\gamma = \nu_p + j\omega_p$ and $\mathrm{Re}[\tilde{\chi}_p(t)] = \bar{\chi}_p(t)$.

From (2.25), we have

$$\Delta\tilde{\chi}_p^0 = \frac{2j\Delta\varepsilon_p\omega_p e^{j\rho_p}}{\gamma}\left(1 - e^{-\gamma\Delta t}\right)^2 \tag{2.26}$$

We finally obtain the recursive expression for the CP term as

$$\boldsymbol{\phi}_p^n = \frac{\boldsymbol{E}^n - \boldsymbol{E}^{n-1}}{2}\Delta\tilde{\chi}^0 + e^{-\gamma\Delta t}\boldsymbol{\phi}_p^{n-1}. \tag{2.27}$$

The calculation for the Drude term is the same as (2.21). The implementation of the above D-CP model can be done with a slight modification to the existing algorithm of the DL model based on the TRC technique. The D-CP equation is explicitly solved in conjunction with the standard finite-difference equations for the $\boldsymbol{H}$ field.

The relation among the RC-related techniques is shown in Fig. 2.1. As can be see, the area of the RC technique is underestimated, since it is calculated using only E^n. For the PLRC technique, the area is more accurate due to its piecewise approximation, but the technique needs two convolution calculations. On the other hand, the area of the TRC technique is identical to the PLRC one, with only one convolution calculation.

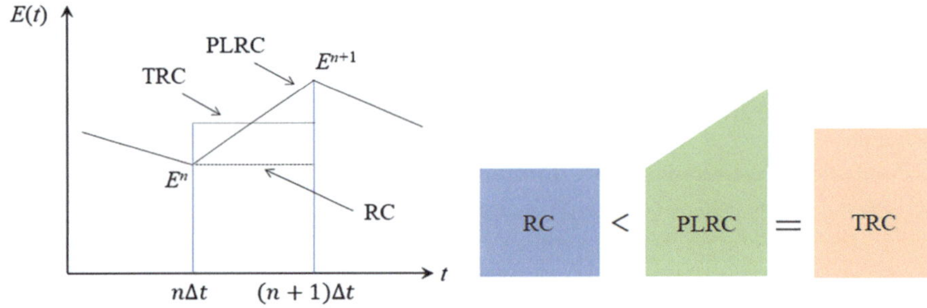

Fig. 2.1 Relation among the RC-related techniques

2.6 Application to Implicit FDTD Method

The above-mentioned techniques have been extended to the implicit LOD-FDTD method [11–16]. Several example of the equations is shown in this section.

For the ADE technique [11, 15], the application of the LOD scheme for the Drude–Lorentz model leads to the equations for the first step as

$$\varepsilon_\infty \frac{E_z^{n+1} - E_z^n}{\Delta t/2} + J_z^{n+1} + J_z^n + \sum_{p=1}^{P} \frac{Q_{p,z}^{n+1} - Q_{p,z}^n}{\Delta t/2} = c\left(\frac{\partial H_y'}{\partial x} + \frac{\partial H_y^{n-1/2}}{\partial x}\right) \tag{2.28a}$$

$$\frac{H_y' - H_y^{n-1/2}}{\Delta t/2} = c\left(\frac{\partial E_z^{n+1}}{\partial x} + \frac{\partial E_z^n}{\partial x}\right) \tag{2.28b}$$

$$\frac{J_z^{n+1} - J_z^n}{\Delta t} + \nu_\mathrm{D} \frac{J_z^{n+1} + J_z^n}{2} = \omega_\mathrm{D}^2 \frac{E_z^{n+1} + E_z^n}{2} \tag{2.28c}$$

$$\frac{Q_{p,z}^{n+1} - 2Q_{p,z}^n - Q_{p,z}^{n-1}}{\Delta t^2} + \nu_p \frac{Q_{p,z}^{n+1} - Q_{p,z}^{n-1}}{2\Delta t} + \omega_p^2 \frac{Q_{p,z}^{n+1} + Q_{p,z}^{n-1}}{2} = \Delta\varepsilon G_p \omega_p^2 \frac{E_z^{n+1} + E_z^n}{2}. \tag{2.28d}$$

For the second step, we have

$$\varepsilon_\infty \frac{E_x^{n+1} - E_x^n}{\Delta t/2} + J_x^{n+1} + J_x^n + \sum_{p=1}^{P} \frac{Q_{p,x}^{n+1} - Q_{p,x}^n}{\Delta t/2} = -c\left(\frac{\partial H_y^{n+1/2}}{\partial z} + \frac{\partial H_y'}{\partial z}\right) \tag{2.29a}$$

$$\frac{H_y^{n+1/2} - H_y'}{\Delta t/2} = -c\left(\frac{\partial E_x^{n+1}}{\partial z} + \frac{\partial E_x^n}{\partial z}\right) \tag{2.29b}$$

$$\frac{J_x^{n+1} - J_x^n}{\Delta t} + \nu_{\mathrm{D}}\frac{J_x^{n+1} + J_x^n}{2} = \omega_{\mathrm{D}}^2\frac{E_x^{n+1} + E_x^n}{2} \tag{2.29c}$$

$$\frac{Q_{p,x}^{n+1} - 2Q_{p,x}^n - Q_{p,x}^{n-1}}{\Delta t^2} + \nu_p\frac{Q_{p,x}^{n+1} - Q_{p,x}^{n-1}}{2\Delta t} + \omega_p^2\frac{Q_{p,x}^{n+1} + Q_{p,x}^{n-1}}{2} = \Delta\varepsilon G_p\omega_p^2\frac{E_x^{n+1} + E_x^n}{2}. \tag{2.29d}$$

To solve the above equations, we first substitute (2.28b)–(2.28d) into (2.28a) and implicitly solve E_z^{n+1}. Once E_z^{n+1} is obtained, the other components are solved explicitly. The solutions for the second step are calculated similarly.

For the Z transform technique [13, 15], (2.9)–(2.11) are solved with the following standard LOD-FDTD equations for the Drude–Lorentz model as

$$\frac{D_z^{n+1} - D_z^n}{\Delta t/2} = c\left(\frac{\partial H_y'}{\partial x} + \frac{\partial H_y^{n-1/2}}{\partial x}\right) \tag{2.30}$$

$$\frac{H_y' - H_y^{n-1/2}}{\Delta t/2} = c\left(\frac{\partial E_z^{n+1}}{\partial x} + \frac{\partial E_z^n}{\partial x}\right) \tag{2.31}$$

for the first step. Using only the z components in (2.9)–(2.11), we substitute (2.9) and (2.31) into (2.30) and implicitly solve the resultant tridiagonal equations, and then explicitly solve (2.9)–(2.11) and (2.31). Note that the relation $E_z^{n+1} = E_z'$ or $D_z^{n+1} = D_z'$, as will be shown in (2.34a), is used in the above implementation. The equations for the second step are

$$\frac{D_x^{n+1} - D_x^n}{\Delta t/2} = -c\left(\frac{\partial H_y^{n+1/2}}{\partial z} + \frac{\partial H_y'}{\partial z}\right) \tag{2.32}$$

$$\frac{H_y^{n+1/2} - H_y'}{\Delta t/2} = -c\left(\frac{\partial E_x^{n+1}}{\partial z} + \frac{\partial E_x^n}{\partial z}\right) \tag{2.33}$$

where the relation $E_x' = E_z^n$ or $D_x' = D_x^n$, as shown in (2.35a), is used, which are similarly solved using only the x components in (2.9)–(2.11).

For the RC-based techniques, we show the equations of the TRC-based LOD-FDTD method for the Drude–Lorentz model as an example [14, 15]. With the TRC technique, we obtain the following equation of the LOD-FDTD for the TM wave (the normalized expression of field components is used):

$$E_x' = E_x^n \tag{2.34a}$$

$$E_z' = \frac{\varepsilon_\infty - \chi^0/2}{\varepsilon_\infty + \chi^0/2}E_z^n + \frac{1}{\varepsilon_\infty + \chi^0/2}\phi_z^n$$

$$+\frac{c\Delta t}{2\left(\varepsilon_{\infty}+\chi^{0}/2\right)}\left(\frac{\partial H_{y}^{\prime}}{\partial x}+\frac{\partial H_{y}^{n-1/2}}{\partial x}\right) \tag{2.34b}$$

$$\frac{H_{y}^{\prime}-H_{y}^{n-1/2}}{\Delta t/2}=c\left(\frac{\partial E_{z}^{\prime}}{\partial x}+\frac{\partial E_{z}^{n}}{\partial x}\right) \tag{2.34c}$$

for the first step and

$$E_{z}^{n+1}=E_{z}^{\prime} \tag{2.35a}$$

$$E_{x}^{n+1}=\frac{\varepsilon_{\infty}-\chi^{0}/2}{\varepsilon_{\infty}+\chi^{0}/2}E_{x}^{\prime}+\frac{1}{\varepsilon_{\infty}+\chi^{0}/2}\phi_{x}^{n}$$

$$-\frac{c\Delta t}{2\left(\varepsilon_{\infty}+\chi^{0}/2\right)}\left(\frac{\partial H_{y}^{n+1/2}}{\partial z}+\frac{\partial H_{y}^{\prime}}{\partial z}\right) \tag{2.35b}$$

$$\frac{H_{y}^{n+1/2}-H_{y}^{\prime}}{\Delta t/2}=-c\left(\frac{\partial E_{x}^{n+1}}{\partial z}+\frac{\partial E_{x}^{\prime}}{\partial z}\right) \tag{2.35c}$$

for the second step, where E' and H' represent the intermediate fields, c is the speed of light in a vacuum. In the first step, we substitute (2.34c) into (2.34b) and implicitly solve the resultant equation. Then, (2.34c) is explicitly solved. In the second step, the equations are calculated in the same way as in the first step. It should be noted that for the LOD-FDTD two implicit and two explicit equations are solved. For the frequency-dependent ADI-FDTD method, refer to references [17–21].

2.7 Comparative Study

2.7.1 Drude Model

Here, we investigate the numerical accuracy of each calculation method for the analysis of a metal-insulator-metal (MIM) waveguide [22–25]. Figure 2.2 illustrates the configuration of the waveguide, the metal of which is chosen to be Ag. The dispersion is expressed with the Drude model, in which $\varepsilon_{\infty}=3.7$, $\omega_{\mathrm{D}}=9.1$ eV, and $\nu_{\mathrm{D}}=0.018$ eV are used in (2.1). The sampling widths are $\Delta x=0.005$ μm and $\Delta z=0.004$ μm. The upper limit of the CFL condition is defined as Δt_{CFL} (= 0.0102 fs), and the ratio of $\Delta t/\Delta t_{\mathrm{CFL}}$ as CFL number (CFLN). The TM pulse wave is launched at the incidence plane (see Fig. 2.2), which consists of the eigenmode field in the x direction and the Gaussian profile with a $1/e$ full-width of 1 μm in the z direction. The pulse is illuminated using the one-way excitation scheme, so that no field propagates toward the -z direction.

Figure 2.3 shows transmission coefficient as a function of CFLN for the several frequency-dependent formulations, in which the explicit and LOD-FDTD methods are used for CFLN ≤ 1 and CFLN ≥ 1, respectively [15]. It is found that the coefficient

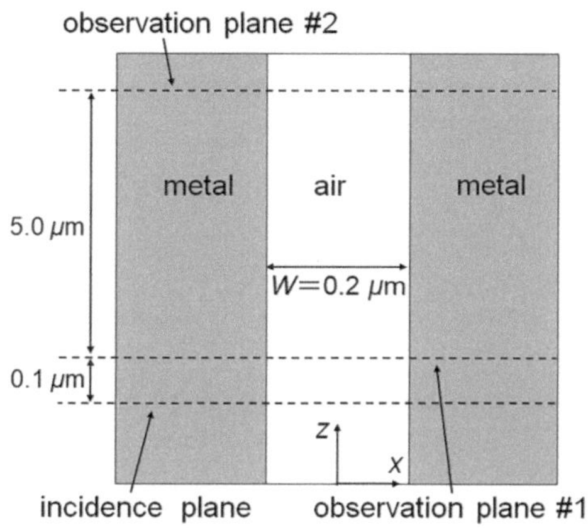

Fig. 2.2 Configuration of a metal cladding optical waveguide. ©[2010] IEEE. Reproduced, with permission, from [15]

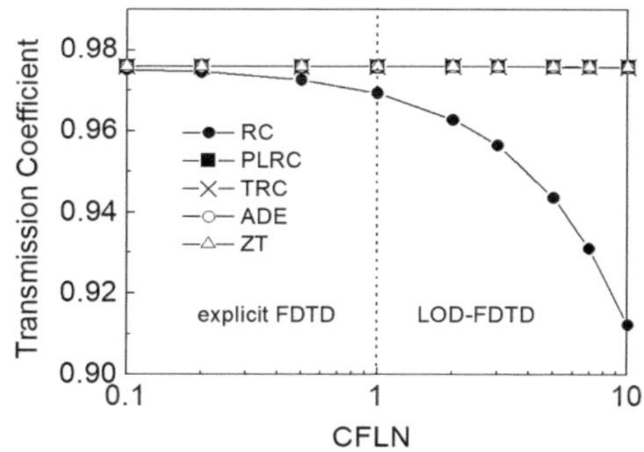

Fig. 2.3 Transmission coefficient versus CFLN for the Ag cladding with the Drude model. ©[2010] IEEE. Reproduced, with permission, from [15]

from the PLRC, TRC, ADE and ZT keeps constant even with a large CFLN up to 10. In contrast, the transmission coefficient from the RC technique becomes small with an increase in CFLN. This is because the former 4 techniques are second-order accurate but the latter one is first-order accurate. Note, at CFLN = 1, that the results obtained from the LOD-FDTD methods are identical to the explicit results, demonstrating the validity of the implicit FDTD methods.

Figure 2.4 depicts the CPU time of each frequency-dependent technique, in which CFLN = 10 is used for the LOD-FDTD method. A PC with Core^{TM}2 Quad processor (2.66 GHz) is used. The results of the RC technique is not shown because of unacceptable accuracy shown in Fig. 2.3. It is found that the CPU time of the LOD-FDTD methods is reduced to less than 30% of that of the explicit counterpart. The implicit FDTD method is shown to be a powerful technique for efficiently analyzing plasmonic devices.

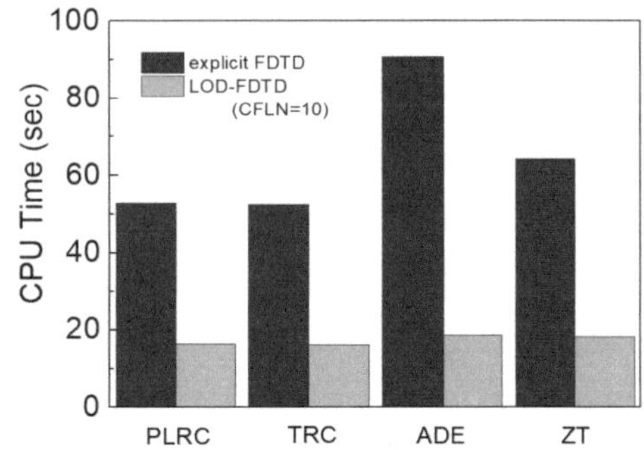

Fig. 2.4 CPU time for the Drude model. ©[2010] IEEE. Reproduced, with permission, from [15]

2.7.2 *Drude–Lorentz Model*

In the second example, Au in the visible region is investigated [15]. The dispersion of Au is accurately described with the Drude–Lorentz model. This is motivated by the fact that Au is frequently applied to the thin metal layer of the surface plasmon resonance sensor operated in the visible region. The parameters are $\varepsilon_\infty = 5.9673$, $\omega_D/2\pi = 2113.6$ THz, $\omega_1/2\pi = 650.07$ THz, $\nu_D/2\pi = 15.92$ THz, $\nu_1/2\pi = 104.86$ THz, and $\Delta\varepsilon = 1.09$ [1], which corresponds to $P = 1$, so that $G_1 = 1$ in (1).

The transmission coefficient versus CFLN at a center wavelength of 0.7 μm is shown in Fig. 2.5. For the explicit FDTDs (CFLN ≤ 1), no difference is found in accuracy among the PLRC, TRC, ADE, and ZT techniques. However, the difference can be seen with the LOD-FDTD being used for CFLN ≥ 1. To examine the accuracy in more detail, the expanded figure for the LOD-FDTD in is shown in Fig. 2.5b. For reference, also included are the eigenmode solution (0.76451) and the value with 0.5% error from the solution, both of which are indicated by the dashed lines. It can be seen that the ZT method achieves higher accuracy by deriving analytical equations to which the finite difference approximation is not applied. The accuracy of the PLRC and TRC methods is almost identical, but careful examination shows that the former is slightly better than the latter. This may be due to the fact that in the former, the electric fields of two consecutive sampling times plus up to $Deltat$ are evaluated using integration, whereas in the latter, only the electric fields of two consecutive sampling times are evaluated. The slight decrease in the accuracy of the ADE method for large CFLN is thought to be due to a decrease in the accuracy of the finite-difference approximation of the second derivative in time in (2.7).

Here, we discuss the computational efficiency of LOD-FDTD when an error of 0.5% is allowed in the transmission coefficients. In this case, the CFLN can be scaled up to 6.2, 5.8, 4.6, and 10 for the PLRC, TRC, ADE, and ZT methods, respectively, as shown in Fig. 2.5b. Figure 2.6 shows the CPU time for each method, indicating that ZT-LOD-FDTD is slightly faster than the other methods due to its larger CFLN. Compared to explicit FDTD, the CPU time for LOD-FDTD is reduced to 48%, 52%, 40%, and 25% for PLRC, TRC, ADE, and ZT methods, respectively. Figure 2.7 shows

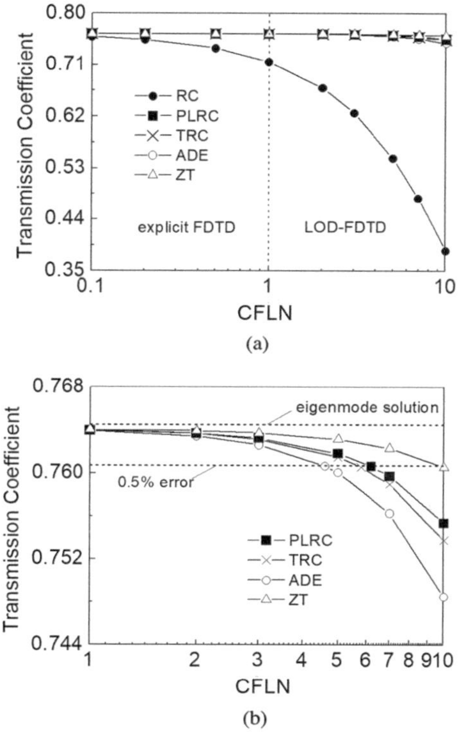

Fig. 2.5 Transmission coefficient versus CFLN for the Au cladding with the Drude–Lorentz model. ©[2010] IEEE. Reproduced, with permission, from [15]

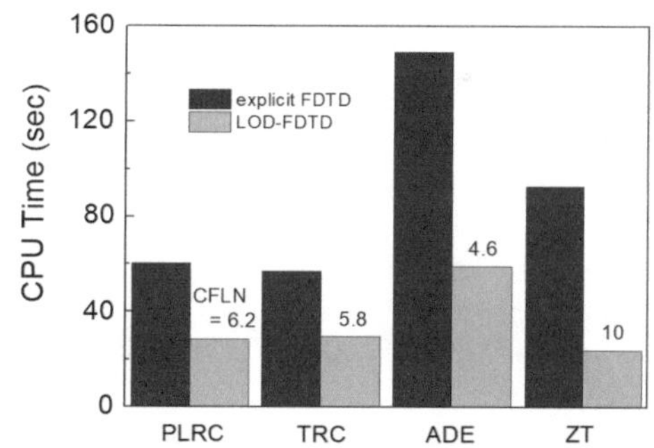

Fig. 2.6 CPU time for the Drude–Lorentz model. ©[2010] IEEE. Reproduced, with permission, from [15]

the required memory, which varies little among the PLRC, TRC, and ADE methods because of the two time step field values required. As in Fig. 2.4, ZT requires slightly larger memory than the other methods.

As seen in [1], for Au in the visible region, there is a strong dispersion in the imaginary part of the dielectric constant, which is taken into account by the Lorentz term. Therefore, large CFLNs cannot be used unless the effect of this strong dispersion is accurately incorporated into the equations to be solved. In this regard, the ZT method is expected to be able to use relatively large CFLNs because the equations

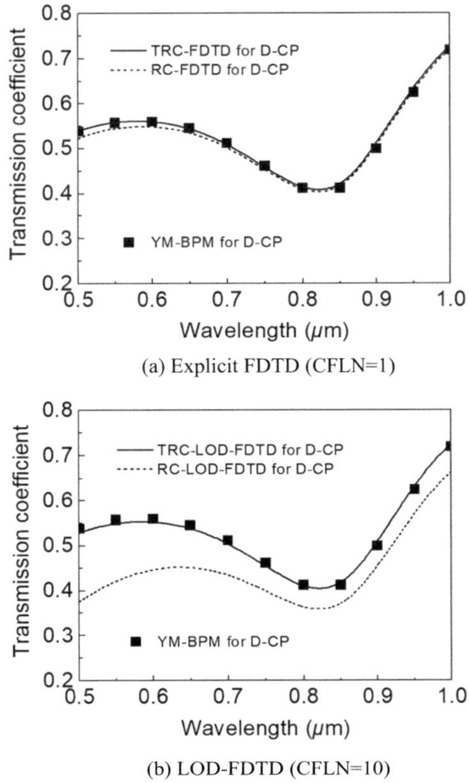

Fig. 2.7 Transmission coefficient versus CFLN for the Al cladding with the D-CP model. Copyright (c)2012 IEICE from [16]

that take dispersion into account can be derived analytically and accurately without using finite-difference approximations.

2.7.3 Drude-Critical Point Model

In the third example, Al in the visible region is investigated. The dispersion of Al is accurately described with the D-CP model, the parameters of which are given in Table 2.1. The transmission coefficients are examined in Fig. 2.7a and b for the explicit FDTD and LOD-FDTD, respectively, comparing the results of the TRC technique with those of the RC technique [16]. For reference, the coefficient from the eigenmode solver (imaginary-distance yee-mesh-based beam-propagation method (YM-BPM) [26]) using the permittivity determined with the D-CP model is included.

In Fig. 2.7a where CFLN $= 1$ is used, the result of the TRC technique for the D-CP model agrees well with the predicted eigenmode results. Although the RC result follows the TRC one, the CFLN of the RC-FDTD should be chosen quite small to achieve the accuracy comparable to that of the TRC-FDTD. More importantly, in

Table 2.1 Parameters for the D-CP model for aluminum [2]

ϵ_∞	1.0
ω_{D} (rad/s)	2.0951×10^{16}
ν_{D} (rad/s)	2.1906×10^{14}
$\Delta\epsilon_p$	6.8011
ω_1 (rad/s)	2.2709×10^{15}
ν_1 (rad/s)	3.7666×10^{14}
ρ_1 (rad)	-0.4780

Fig. 2.7b, the TRC technique maintains the accuracy even for CFLN $= 10$, while the RC results deviate from the eigenmode results. In this case, the computational time of the LOD-FDTD for CFLN $= 10$ is reduced to only 25% of that of the explicit FDTD for CFLN $= 1$, using a PC with an Intel Core2 Extreme processor (2.20 GHz).

Finally, we numerically evaluate the order of the TRC and RC techniques, paying attention to CFLN. Since the convergence of the numerical results is not complete, we determine a more accurate transmission coefficient expected at CFLN $= 0$ using an extrapolation technique. Figure 2.8 shows the deviation of the transmission coefficient relative to the extrapolated value at a center wavelength of $\lambda = 0.75$ μm as a function of CFLN. The PLRC results are also included for comparison. Taking advantage of the unconditional stability, we use the LOD FDTD for CFLN ≥ 1. For CFLN ≤ 1, the traditional explicit FDTD is used. For CFLN $= 1$, the identical results are obtained from both the LOD-FDTD and the explicit FDTD, validating the LOD-FDTD. As expected, the RC technique is found to be first order accurate. In contrast, the TRC and PLRC techniques exhibit second order accuracy for CFLN

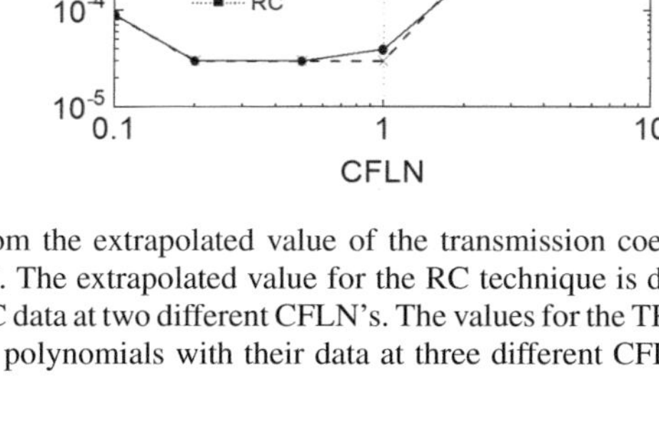

Fig. 2.8 Deviation from the extrapolated value of the transmission coefficient at $\lambda = 0.75$ μm as a function of CFLN. The extrapolated value for the RC technique is derived from a first-order polynomial with the RC data at two different CFLN's. The values for the TRC and PLRC techniques are from second-order polynomials with their data at three different CFLN's. Copyright (c)2012 IEICE from [16]

≥ 1, while for CFLN ≤ 1 the results reach the accuracy limit of this analysis. As a result, the TRC technique is found to be as accurate as the PLRC technique, even with the single convolution integral calculation.

2.8 Concluding Remarks

Several frequency-dependent FDTD formulations have been shown for the analysis of dispersive media. Three frequently-used techniques are explained such as the ADE, RC-related and Z-transform (ZT) techniques. Application of the frequency-dependent FDTD method to implicit calculations is also presented. The accuracy of these techniques is compared and discussed for an MIM waveguide. The metal of the waveguide is selected to be Ag, Au and Al, the dispersions of which are expressed with the Drude, Drude–Lorentz and Drude-Critical Point models, respectively. For Ag with the Drude model, the use of the simple TRC-LOD-FDTD is preferable, due to its acceptable accuracy, less CPU time, and moderate memory requirement. For Au with the Drude–Lorentz model, the ZT technique provides better accuracy since the resultant equation is derived analytically and accurately without using finite-difference approximations.

References

1. A. Vial, A.-S. Grimault, D. Macías, D. Barchiesi, M.L. de la Chapelle, Improved analytical fit of gold dispersion: application to the modeling of extinction spectra with a finite-difference time-domain method. Phys. Rev. B **71**, 085416-1-7 (2005)
2. A. Vial, Implementation of the critical points model in the recursive convolution method for modelling dispersive media with the finite-difference time domain method. J. Opt. A-Pure Appl. Opt. **9**, 745–748 (2007)
3. A. Vial, T. Laroche, Description of dispersion properties of metals by means of the critical points model and application to the study of resonant structures using the FDTD method. J. Phys. D-Appl. Phys. **40**(22), 7152–7158 (2007)
4. T. Kashiwa, I. Fukai, A treatment by FD-TD method of dispersive characteristics associated with electronic polarization. Microw. Opt. Technol. Lett. **3**(6), 203–205 (1990)
5. R.M. Joseph, S.C. Hagness, A. Taflove, Direct time integration of Maxwell equations in linear dispersive media with absorption for scattering and propagation of femtosecond electromagnetic pulses. Opt. Lett. **16**(18), 1412–1414 (1991)
6. D.M. Sullivan, Frequency-dependent FDTD methods using *Z* Transforms. IEEE Trans. Antennas Propag. **40**(10), 1223–1230 (1992)
7. R.J. Luebbers, F. Hunsberger, FDTD for *N*th-order dispersive media. IEEE Trans. Antennas Propag. **40**(11), 1297–1301 (1992)
8. D.F. Kelley, R.J. Luebbers, Piecewise linear recursive convolution for dispersive media using FDTD. IEEE Trans. Antennas Propag. **44**(6), 792–797 (1996)
9. R. Siushansian, J. LoVetri, A comparison of numerical techniques for modeling electromagnetic dispersive media. IEEE Microw. Guided Wave Lett. **5**(12), 426–428 (1995)

10. R. Siushansian, J. LoVetri, Efficient evaluation of convolution integrals arising in FDTD formulations of electromagnetic dispersive media. J. Electromag. Waves Appli. **11**(1), 101–117 (1997)
11. J. Shibayama, R. Takahashi, J. Yamauchi, H. Nakano, Frequency-dependent LOD-FDTD implementations for dispersive media. Electron. Lett. **42**(19), 1084–1086 (2006)
12. J. Shibayama, R. Takahashi, J. Yamauchi, H. Nakano, Frequency-dependent locally one-dimensional FDTD implementation with a combined dispersion model for the analysis of surface plasmon waveguides. IEEE Photon. Technol. Lett. **20**(10), 824–826 (2008)
13. J. Shibayama, R. Takahashi, A. Nomura, J. Yamauchi, H. Nakano, Concise frequency-dependent formulation for LOD-FDTD method using *Z* transforms. Electron. Lett. **44**(16), 949–950 (2008)
14. J. Shibayama, R. Ando, A. Nomura, J. Yamauchi, H. Nakano, Simple trapezoidal recursive convolution technique for the frequency-dependent FDTD analysis of a Drude-Lorentz model. IEEE Photon. Technol. Lett. **21**(2), 100–102 (2009)
15. J. Shibayama, A. Nomura, R. Ando, J. Yamauchi, H. Nakano, A frequency-dependent LOD-FDTD method and its application to the analyses of plasmonic waveguide devices. IEEE J. Quantum Electron. **46**(1), 40–49 (2010)
16. J. Shibayama, K. Watanabe, R. Ando, J. Yamauchi, H. Nakano, Frequency-dependent formulations of a Drude-critical points model for explicit and implicit FDTD methods using the trapezoidal RC technique. IEICE Trans. Electron. **E95-C**(4), 725–732 (2012)
17. L. Xu, N. Yuan, PLJERC-ADI-FDTD method for isotropic plasma. IEEE Microw. Wirel. Compon. Lett. **15**(4), 277–279 (2005)
18. K.-Y. Jung, F.L. Teixeira, Multispecies ADI-FDTD algorithm for nanoscale three-dimensional photonic metallic structures. IEEE Photon. Technol. Lett. **19**(8), 586–588 (2007)
19. H.K. Rouf, F. Costen, S.G. Garcia, Reduction of numerical errors in frequency dependent ADI-FDTD. Electron. Lett. **46**(7), 489–490 (2010)
20. J.A. Pereda, A. Grande, O. Gonzalez, A. Vegas, An efficient 2D ADI-FDTD modeling of high-order, frequency-dependent media. IEEE Antennas Wirel. Propagat. Lett. **8**, 724–727 (2009)
21. K.P. Prokopidis, D.C. Zografopoulos, Modeling of dispersive media in ADI-FDTD method with complex-conjugate pole residue pairs. J. Opt. Soc. Am. B **39**(1), 273–285 (2022)
22. K. Tanaka, M. Tanaka, Simulations of nanometric optical circuits based on surface plasmon polariton gap waveguide. Appl. Phys. Lett. **82**(8), 1158–1160 (2003)
23. J. Takahara, F. Kusunoki, Guiding and nanofocusing of two-dimensional optical beam for nanooptical integrated circuits. IEICE Trans. Electron. **E90-C**(1), 87–94 (2007)
24. A. Hossieni, Y. Massoud, A low-loss metal-insulator-metal plasmonic bragg reflector. Opt. Express **14**(23), 11318–11323 (2006)
25. Z. Han, E. Forsberg, S. He, Surface plasmon Bragg gratings formed in metal-insulator-metal waveguides. IEEE Photon. Technol. Lett. **19**(2), 91–93 (2007)
26. J. Yamauchi, T. Mugita, H. Nakano, Implicit Yee-mesh-based finite-difference full-vectorial beam-propagation method. J. Lightw. Technol. **23**(5), 1947–1955 (2005)

Chapter 3
Plasmonic Waveguides Devices

Abstract This chapter presents some plasmonic waveguides devices. First, a plasmonic grating made of a two-dimensional (2-D) metal-insulator-metal (MIM) waveguide is analyzed with attention to its propagation characteristics. It is shown that the grating acts as a band stop filter with relatively large sidelobes. To reduce the sidelobes, an apodized grating is introduced into the filter. It is found that the sidelobes are effectively suppressed, leading to a step like band stop property. In addition, a plasmonic microcavity is analyzed, in which a defect section is introduced into the band-stop filter. Next, the performance of MIM waveguide-based band-pass filters is investigated, in which the results from 2-D model are compared with those from three-dimensional (3-D) one. It is revealed that the performance from the 2-D waveguides cannot be obtained by simply adopting the 3-D ones. Finally, the rod waveguides are efficiently analyzed using the cylindrical and body-of-revolution (BOR) FDTD methods. The formulation of the LOD-BOR-FDTD method is given and the reflection characteristics of the rod waveguide are presented.

3.1 Introduction

Plasmonic waveguide devices consisting of metal-insulator-metal (MIM) waveguides have been widely studied because the sizes of the devices can be significantly reduced compared to those of the conventional waveguide devices made only of insulator materials. So far, many MIM-based plasmonic devices have been theoretically investigated [1–10], such as waveguides, filters, resonators, splitters, modulators, and so on. In this chapter, we first analyze band-stop filters made of MIM gratings [2]. It is shown that by using apodized gratings, band-stop characteristics can be obtained in which sidelobes can be successfully reduced. The band-stop filter can also be used as a microcavity with a defect section. A transmission peak is obtained in the stop-band characteristics. Next, we study band-pass filters made of MIM gratings, paying attention to the dimension of the device [3]. A two-dimensional (2-D) filter is shown to have an excellent band-pass property, while a three-dimensional (3-D) filter cannot reproduce the results similar to its 2-D counterpart when the metal thickness is thin. If we want to have an excellent band-pass property with a 3-D filter,

J. Shibayama, *FDTD Analysis of Plasmonic Devices*, Springer Series in Optical Sciences 254, https://doi.org/10.1007/978-981-96-8130-3_3

its thickness should be chosen to be more than 3 μm. Final examination is given to band reflection filters using a metal rod waveguide grating. The formulation of a frequency-dependent locally one-dimensional (LOD) FDTD method is provided in the body-of-revolution (BOR) technique [11–13] with fundamental scheme. The reflection characteristics are shown to agree well with the traditional explicit results, while the computational time is reduced to about 50% of the explicit counterpart.

3.2 Band-Stop Filter Using MIM Waveguide

We here investigate the characteristics of a band-stop plasmonic filter composed of MIM waveguides [1, 2]. The structures are shown in Fig. 3.1, in which (a) conventional grating and (b) apodized grating. Note that Ag is used for a metal and its dispersion property is expressed by the Drude model discussed in this chapter. The structural parameters are as follows: $W_1 = 0.1$ μm, $W_2 = 0.2$ μm, $L_P = 0.66$ μm, $L_S = 0.292$ μm, and the number of periods is 19. We excite a pulse wave along the incidence plane using the eigenmode of the MIM waveguide and observe the spectrum at the observation plane.

Figure 3.2 shows the transmission coefficient as a function of wavelength obtained from the LOD-FDTD method based on the TRC technique. It is seen that the band-stop property is obtained around 1.5 μm. Note, however, that the sidelobe can be seen outside of the stop band. To reduce the sidelobe, an apodized grating is introduced. The results are also shown in Fig. 3.2, in which a linear apodization is introduced for the 5 or 9 periods at both input and output ports. It is seen that the sidelobe is successfully reduced, when compared with the filter with normal gratings, leading to a step-like band-stop property.

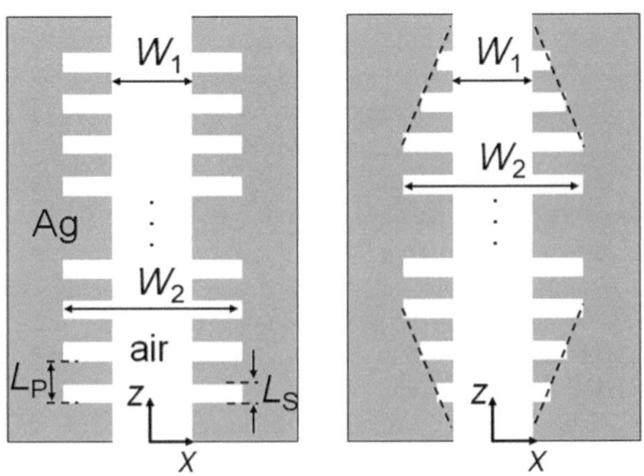

(a) Conventional (b) Apodized

Fig. 3.1 Configuration of band-top filter composed of gratings. © [2010] IEEE. Reproduced, with permission, from [2]

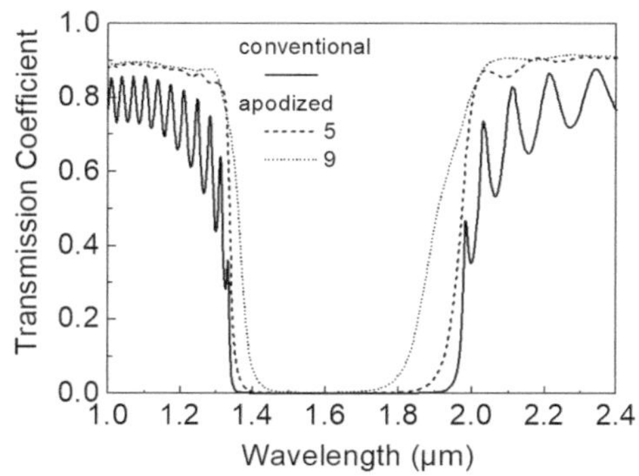

Fig. 3.2 Transmission coefficient for the several gratings. © [2010] IEEE. Reproduced, with permission, from [2]

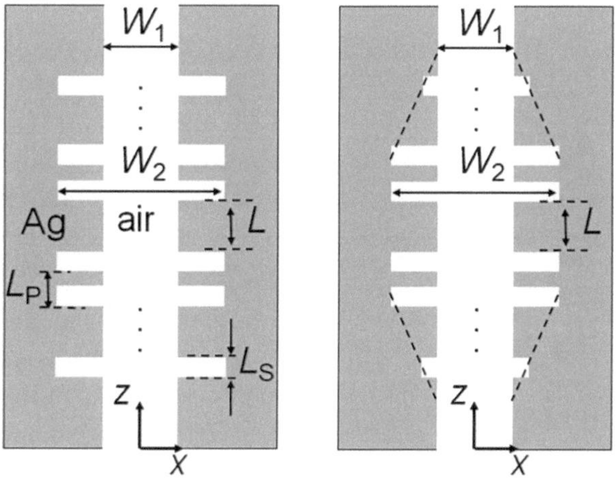

Fig. 3.3 Configuration of plasmonic microcavities composed of gratings. © [2010] IEEE. Reproduced, with permission, from [2]

It is worth mentioning that the band-stop filter can be used as a plasmonic microcavity with a defect section. Figure 3.3 shows the configuration of plasmonic microcavity, in which the defect section is introduced into the center of the grating. The defect is $L = 0.688$ μm in length and the number of gratings is 18. Figure 3.4 depicts the transmission property of the microcavity. It is seen that the transmission peak appears around $\lambda = 1.6$ μm. The results of the apodized microcavity are also included. The peak power increases with an increase in the number of the apodized gratings. Although not illustrated, the peak wavelength can be tuned from 1.5 to 1.8 μm by varying the width of the air core region of the MIM wavegude.

Fig. 3.4 Transmission for the several micro cavities. © [2010] IEEE. Reproduced, with permission, from [2]

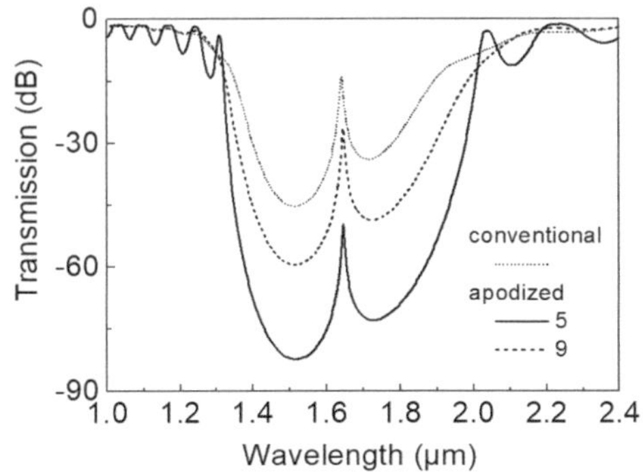

3.3 Band-Pass Filter Using MIM Waveguide

A band-pass filter based on plasmonic grating is investigated with attention to the difference between 2-D and 3-D models [3]. Figure 3.5 shows the gating-type band-pass filter made of Ag. The gap width of the input and output sections is $w = 0.05$ μm. The metal thickness is chosen to be $h = \infty$ for the 2-D filter, and $h = 0.05, 0.1, 0.5, 1$ and 3 μm for the 3-D filters. The number of gratings is three, in which $l = 0.1$ μm and $d = 0.55$ μm.

Figure 3.6 depicts the power transmission of the filter. It is found that an excellent band-pass property can be obtained with the 2-D filter, in which more than 50% transmission is achieved from 1.14 to 1.44 μm with a peak transmission of more than 90%. This pass band appears between the stop bands arising from a large difference of the effective indexes of the convex and concave sections in the gratings. On the other hand, the 3-D filter yields a property with less output power, especially for a thin metal thickness. It should be noted that the band-pass property by the 2-D filter cannot be reproduced even by the 3-D one with a thick metal thickness.

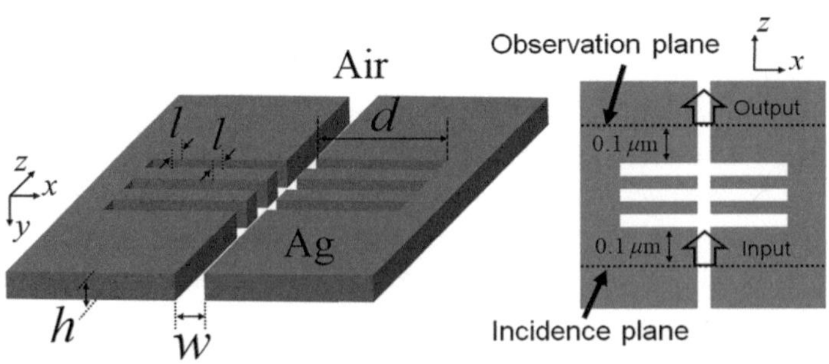

Fig. 3.5 Configuration of band-pass filter composed of gratings. Copyright © 2016 IEICE from [3]

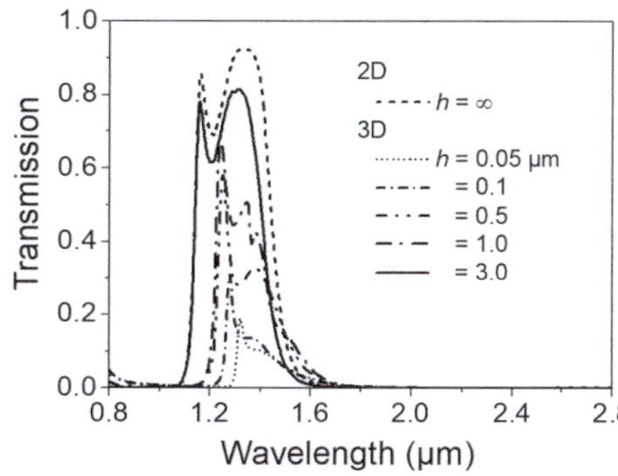

Fig. 3.6 Transmission. Copyright © 2016 IEICE from [3]

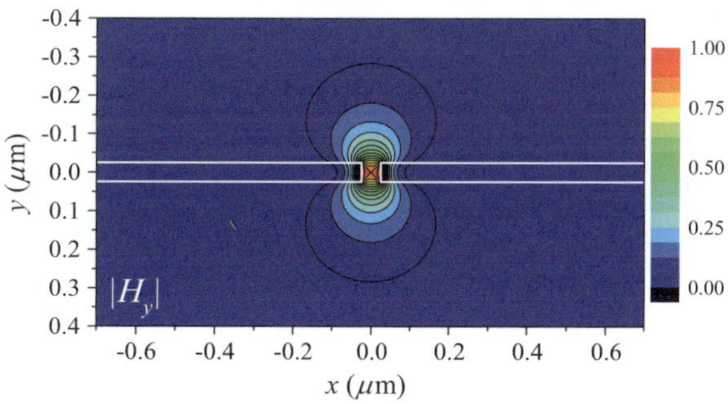

Fig. 3.7 Eigenmode field at $\lambda = 1.32$ μm in the x-y plane. Copyright © 2016 IEICE from [3]

To investigate the reason for the reduced transmission of the 3-D filter, we illustrate the field distributions in the x-y plane for $h = 0.05$ μm, the thickness of which is frequently used for a plasmonic waveguide. Figure 3.7 shows the eigenmode filed of the gap waveguide used to excite the input pulse. It can be seen that the field extends into the air region, reducing the effect of the grating (the field in the 2D filter is completely confined to the waveguide and gratings). Furthermore, we illustrate the propagating fields obtained from the FDTD analysis. Figure 3.8a and b respectively depict the propagating fields in the x-y and y-z planes. Appreciable radiation occurs and the fields deviate from the eigenmode filed shown in Fig. 3.7. This stems from the large discontinuities in the grating section, reducing the coupling efficiency of the fields between the input and grating sections. As a result, it is difficult for the 3-D grating-type filter with a thin metal layer to obtain a high transmission.

3.4 Reflection Filter Using Rod Waveguide

To efficiently analyze a plasmonic rod waveguide, we first present the equations of the body-of-revolution (BOR) FDTD method based on the locally one-dimensional (LOD) and fundamental schemes.

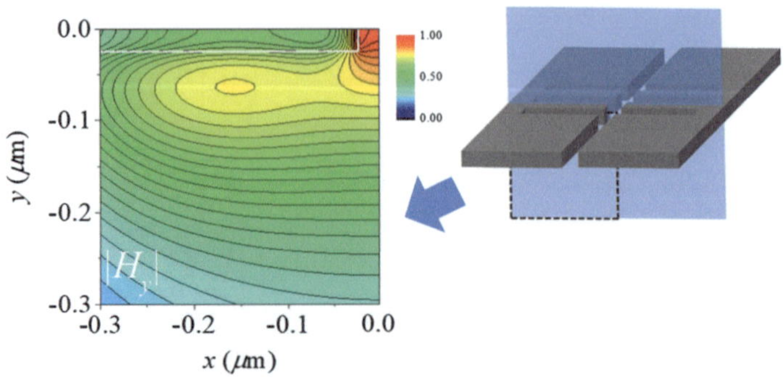

(a) x-y plane

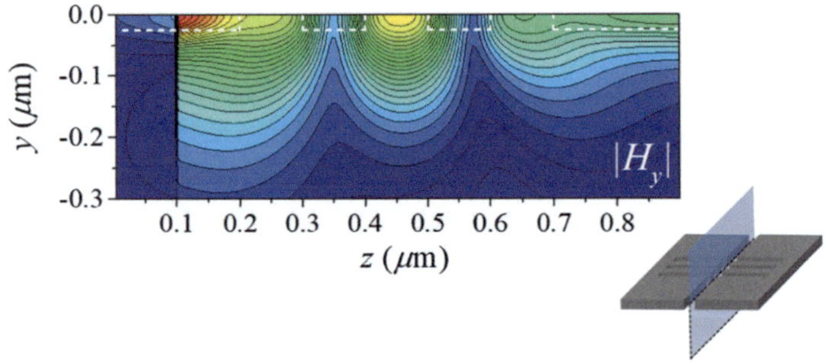

(b) y-z plane

Fig. 3.8 Cross-sectional view of the field distribution

The matrix form of the frequency-dependent LOD-BOR-FDTD is obtained with the TRC technique discussed in Chap. 3, in which the term $\boldsymbol{d}$ considering dispersion characteristics is added to the frequency-independent formulation in [12]. The matrix form for the Drude-Lorentz model is derived as (without the fundamental scheme)

$$\left([I] - \frac{\Delta t}{2}[A]\right) \boldsymbol{u}^{n+1/2} = \left([I] + \frac{\Delta t}{2}[A]\right) \boldsymbol{u}^{n} + \boldsymbol{d}^{n} \tag{3.1}$$

for the first step and

$$\left([I] - \frac{\Delta t}{2}[B]\right) \boldsymbol{u}^{n+1} = \left([I] + \frac{\Delta t}{2}[B]\right) \boldsymbol{u}^{n+1/2} + \boldsymbol{d}^{n+1/2} \tag{3.2}$$

for the second step, where

$$[A] = \begin{bmatrix} 0 & 0 & 0 & 0 & -\frac{\gamma c}{\epsilon_r}\frac{\partial}{\partial z} & 0 \\ 0 & 0 & 0 & 0 & 0 & -\frac{\alpha c}{\epsilon_r}\frac{\partial}{\partial \rho} \\ 0 & 0 & 0 & -\frac{mc}{\epsilon_r \rho} & 0 & 0 \\ 0 & 0 & \frac{mc}{\mu_r \rho} & 0 & 0 & 0 \\ -\frac{\gamma c}{\mu_r}\frac{\partial}{\partial z} & 0 & 0 & 0 & 0 & 0 \\ 0 & -\frac{\alpha c}{\mu_r \rho}\frac{\partial(\rho)}{\partial \rho} & 0 & 0 & 0 & 0 \end{bmatrix}$$

$$[B] = \begin{bmatrix} 0 & 0 & 0 & 0 & 0 & \frac{mc}{\epsilon_r \rho} \\ 0 & 0 & 0 & \frac{\gamma c}{\epsilon_r}\frac{\partial}{\partial z} & 0 & 0 \\ 0 & 0 & 0 & 0 & \frac{\alpha c}{\epsilon_r \rho}\frac{\partial(\rho)}{\partial \rho} & 0 \\ 0 & \frac{\gamma c}{\mu_r}\frac{\partial}{\partial z} & 0 & 0 & 0 & 0 \\ 0 & 0 & \frac{\alpha c}{\mu_r}\frac{\partial}{\partial \rho} & 0 & 0 & 0 \\ -\frac{mc}{\mu_r \rho} & 0 & 0 & 0 & 0 & 0 \end{bmatrix}$$

$$\boldsymbol{u} = [E_\rho, E_\phi, E_z, H_\rho, H_\phi, H_z]^T$$

$$\boldsymbol{d} = \left[\frac{\phi_\rho - \chi^0 E_\rho}{\varepsilon_\infty + \chi^0/2}, \frac{\phi_\phi - \chi^0 E_\phi}{\varepsilon_\infty + \chi^0/2}, \frac{\phi_z - \chi^0 E_z}{\varepsilon_\infty + \chi^0/2}, 0, 0, 0\right]^T$$

$$\chi^0 = \chi_{\mathrm{D}}^0 + \sum_{p=1}^{P} \mathrm{Re}\left[\chi_p^0\right]$$

$$\phi_\xi^n = \phi_{\xi \mathrm{D}}^n + \sum_{p=1}^{P} \mathrm{Re}\left[\phi_{\xi \mathrm{p}}^{\mathrm{n}}\right]$$

$$\phi_{\xi \mathrm{D}}^n = \frac{E_\xi^n + E_\xi^{n-1/2}}{2} \Delta \chi_{\mathrm{D}}^0 + e^{-\nu_{\mathrm{D}} \Delta t/2} \phi_{\xi \mathrm{D}}^{n-1/2}$$

$$\phi_{\xi p}^n = \frac{E_\xi^n + E_\xi^{n-1/2}}{2} \Delta \chi_p^0 + e^{\gamma \Delta t/2} \phi_{\xi p}^{n-1/2}$$

in which ξ is ρ, ϕ or z, ϵ_r is the relative permittivity, μ_r is the relative permeability, c is the speed of light in a vacuum, and m is the azimuthal mode index. To equalize the field amplitudes between E and H, we employ the normalized expression of the fields. The parameters γ, χ_{D}^0, χ_p^0, $\Delta\chi_{\mathrm{D}}^0$, and $\Delta\chi_p^0$ are derived with the TRC technique for the Drude-Lorentz model and the details can be found in Chap. 2.5.

Here, we introduce the fundamental scheme for the efficient implementation. The equation of (3.1) is rewritten as

$$\left([I] - \frac{\Delta t}{2}[A]\right) \boldsymbol{u}^{n+1/2} = 2\boldsymbol{u}^n - \left([I] - \frac{\Delta t}{2}[A]\right) \boldsymbol{u}^n + \boldsymbol{d}^n$$

$$\left([I] - \frac{\Delta t}{2}[A]\right)\left(\boldsymbol{u}^{n+1/2} + \boldsymbol{u}^n\right) = 2\boldsymbol{u}^n + \boldsymbol{d}^n. \tag{3.3}$$

The second parenthesis in the left-hand side of (3.3) is replaced with the auxiliary variables $\boldsymbol{v} = [e_\rho, e_\phi, e_z, h_\rho, h_\phi, h_z]^T$. Doing so leads to the basic equations of the fundamental LOD-BOR-FDTD as

$$\left([I] - \frac{\Delta t}{2}[A]\right)\boldsymbol{v}^{n+1/2} = 2\boldsymbol{u}^n + \boldsymbol{d}^n \tag{3.4}$$

$$\boldsymbol{u}^{n+1/2} = \boldsymbol{v}^{n+1/2} - \boldsymbol{u}^n. \tag{3.5}$$

First, $\boldsymbol{v}^{n+1/2}$ is implicitly calculated using (3.4) and then $\boldsymbol{u}^{n+1/2}$ is explicitly obtained using (3.5). When compared to (3.1) without the fundamental scheme, (3.4) and (3.5) have no derivative in the right-hand sides. In fact, the number of arithmetic operations of the resultant equations in the right-hand sides is significantly reduced, i.e., the number of multiplication and division operations is 8 and that of addition and subtraction operations is 32, while those for the conventional scheme are respectively 22 and 44 (for the case of $m = 0$).

To investigate the fundamental LOD-BOR-FDTD method, we analyze the plasmonic grating made of Ag, shown in Fig. 3.9. The Drude model is used for Ag discussed in Chap. 1. The spatial sampling widths are chosen to be $\Delta\rho = 0.005$ μm and $\Delta z = 0.004$ μm (the number of sampling points is 500×3000 in the ρ and z directions, respectively). From these parameters, the CFLN limit is calculated to be $\Delta t_{\text{CFL}} \simeq 0.0094$ fs. For the explicit FDTD, we take 13500 time steps, resulting in $\simeq 127$ fs. Figure 3.10 shows the reflection coefficient calculated using the Fourier transform of the incident and reflected fields. It is found that the result of the fundamental LOD scheme for CFLN=7 is exactly the same as that of the conventional LOD-BOR scheme and agrees well with the explicit counterpart.

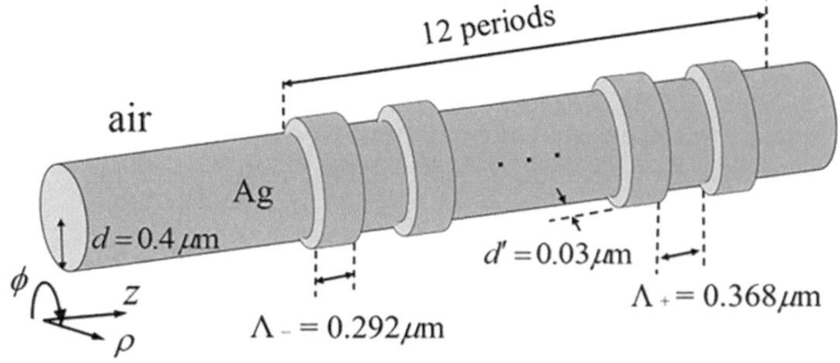

Fig. 3.9 Plasmonic grating. Copyright © 2014 IEICE from [13]

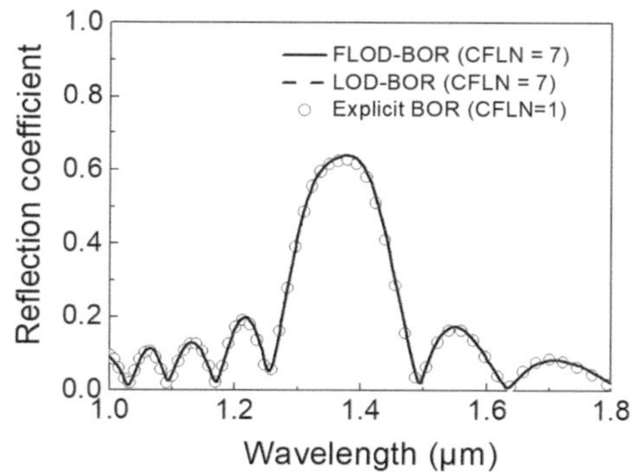

Fig. 3.10 Reflection characteristics obtained from the explicit and LOD-FDTD methods. Copyright © 2014 IEICE from [13]

Finally, we mention the computation time using a PC with Core i7 (2.93 GHz). The explicit and fundamental LOD methods require 243 and 125 min, respectively. As a result, the computation time of the present FLOD-BOR-FDTD is successfully reduced to $\simeq$50% of that of the explicit BOR-FDTD.

3.5 Concluding Remarks

A 2-D MIM waveguide has been analyzed using the frequency-dependent FDTD method. It has been shown that the grating acts as a band-stop filter with relatively large sidelobes. An apodized grating was introduced into the filter. The sidelobes were shown to be effectively suppressed, resulting in a step-like band-stop property. In addition, a plasmonic microcavity with a defect section was analyzed. Next, the performance of MIM waveguide-based bandpass filters was investigated. It was found that the performance of the 2-D waveguides cannot be obtained by simply adopting the 3D waveguides with a reasonable metal thickness. Finally, the rod waveguides were efficiently analyzed using the cylindrical and BOR FDTD methods.

References

1. Z. Han, E. Forsberg, S. He, Surface plasmon Bragg gratings formed in metal-insulator-metal waveguides. IEEE Photon. Technol. Lett. **19**(2), 91–93 (2007)
2. J. Shibayama, A. Nomura, R. Ando, J. Yamauchi, H. Nakano, A frequency-dependent LOD-FDTD method and its application to the analyses of plasmonic waveguide devices. IEEE J. Quantum Electron. **46**(1), 40–49 (2010)
3. J. Shibayama, Y. Wada, J. Yamauchi, H. Nakano, Analysis of two- and three-dimensional plasmonic waveguide band-pass filters using the TRC-FDTD method. IEICE Trans. Electron. **E99-C**(7), 817–819 (2016)
4. S. Naghizadeh, Ş.E. Kocabaş, Guidelines for designing 2D and 3D plasmonic stub resonators. J. Opt. Soc. Am. B **34**, 207–217 (2017)

5. Y. Naka, M. Nishimoto, M. Yokota, Analysis of optical resonator constructed by two-dimensional MDM plasmonic waveguide. IEICE Trans. Electron. **E106-C**(3), 103–106 (2023)
6. Y. Naka, M. Nishimoto, M. Yokota, Analysis of optical power splitter with resonator structure constructed by two-dimensional MDM plasmonic waveguide. IEICE Trans. Electron. **E107-C**(5), 141–145 (2024)
7. Y. Kurokawa, H.T. Miyazaki, Metal-insulator-metal plasmon nanocavities: analysis of optical properties. Phys. Rev. B **75**, 035411-1-13 (2007)
8. P. Neutens, L. Lagae, G. Borghs, P. Van Dorpe, Plasmon filters and resonators in metal-insulator-metal waveguides. Opt. Exp. **20**(4), 3408–3423 (2012)
9. J. Chen, G.A. Smolyakov, S.R.J. Brueck, K.J. Malloy, Surface plasmon modes of finite, planar, metal-insulator-metal plasmonic waveguides. Opt. Exp. **16**(19), 14902–14909 (2008)
10. H. Choo, M.K. Kim, M. Staffaroni et al., Nanofocusing in a metal-insulator-metal gap plasmon waveguide with a three-dimensional linear taper. Nature Photon **6**, 838–844 (2012)
11. J. Shibayama, B. Murakami, J. Yamauchi, H. Nakano, LOD-BOR-FDTD algorithm for efficient analysis of circularly symmetric structures. IEEE Microw. Wirel. Compon. Lett. **19**(2), 56–58 (2009)
12. J. Shibayama, T. Oikawa, J. Yamauchi, H. Nakano, Efficient LOD-BOR-FDTD implementation based on a fundamental scheme. IEEE Photon. Technol. Lett. **24**(11), 957–959 (2012)
13. J. Shibayama, T. Oikawa, T. Hirano, J. Yamauchi, H. Nakano, Fundamental LOD-BOR-FDTD method for the analysis of plasmonic devices. IEICE Trans. Electron. **E97-C**(7), 707–709 (2014)

Chapter 4
Demultiplexer Based on 3-D Metal-Insulator-Metal Waveguide

Abstract This chapter presents a novel functional device based on three-dimensional (3-D) metal-insulator-metal (MIM) waveguides. The performance of the demultiplexer is demonstrated for wavelengths of 1.31 and 1.55 μm. The demultiplexer is composed of slit- and stub-type filters based on metal-insulator-metal waveguides. To achieve a narrower bandwidth for the slit-type filter, a dielectric material is inserted in the slit section and the metal thickness is intentionally set to 0.2 μm. Then, a demultiplexer with two filters in parallel is considered. Numerical results show that the separation between the two output ports is 18.9 dB at 1.31 μm and only 9.8 dB at 1.55 μm. To improve this separation, stub sections forming attenuation poles are introduced into the waveguide. As a result, it is found that a separation of more than 20 dB is obtained at both wavelengths. All calculations are performed using a GPU-accelerated finite-difference time-domain method based on the trapezoidal recursive convolution method.

4.1 Introduction

Surface plasmon polaritons (SPPs) are known to be supported at optical frequencies along metal-insulator interfaces. Waveguide-type plasmonic devices that take advantage of this SPP propagation property are widely used. In particular, plasmonic gap waveguides consisting of metal-insulator-metal (MIM) structures have attracted much attention because they can confine SPPs to sub-wavelength regions [1, 2]. Furthermore, state-of-the-art fabrication techniques make it possible to construct nanoscale devices using MIMs. These facts suggest the possibility of realizing miniature optical circuits.

Many plasmonic devices based on MIM waveguides have been studied theoretically, including resonators [3–6], Bragg reflectors [7–11], filters [12–18], and splitters [19–25]. However, all of these studies have been conducted using two-dimensional (2-D) models, and little attention has been paid to practical three-dimensional (3-D) models. This is because the properties of a 3-D model can be approximately evaluated with a 2-D model as long as the metal layer is at least 1 μm [26] thick. However, the use of thicker metals would inhibit the advantages of

J. Shibayama, *FDTD Analysis of Plasmonic Devices*, Springer Series in Optical Sciences 254, https://doi.org/10.1007/978-981-96-8130-3_4

miniaturization of plasmonic devices. Therefore, 3-D analysis is essential to study the practical application and miniaturization of plasmonic devices with thin metal layers. In this chapter, we investigate the possible structure of a 3-D wavelength demultiplexer consisting of an improved filter based on a MIM waveguide with a thin metal layer [27]. In what follows, calculations are performed with a GPU-accelerated frequency-dependent finite-difference time-domain (FD-FDTD) method based on the trapezoidal recursive convolution technique (TRC) discussed in Chap. 2.

4.2 GPU Acceleration for the Analysis of Plasmonic Devices

The metal used in this paper is chosen to be Ag that is expressed with the Drude model [8]. The skin depth of the metal is a few tens of nanometers. It follows that the spatial sampling meshes should be chosen to be several nanometers. This causes a quite small time step due to the stability condition of the FDTD method, giving rise to a long computation time. We have two possible ways to alleviate this problem: one is to use implicit FDTD methods [28, 29] and the other is to resort to a hardware accelerator [30, 31]. We here choose the latter technique, that is, the GPU-accelerated FD-FDTD method with the convolutional perfect matched layers is implemented with the OpenACC and OpenMP directives [32].

The FD-FDTD method used here is based on the TRC technique in Chap. 2, yielding the second-order accuracy as in the piecewise linear recursive convolution (PLRC) technique [33]. It is worth mentioning that the TRC technique requires only a single convolution calculation, while the PLRC one involves two convolution calculations.

To check the effectiveness of the GPU-FDTD calculations, we analyze a typical 3-D MIM waveguide with a gap width of 0.05 μm and a metal thickness of 0.05 μm [2]. The spatial sampling meshes are chosen to be $\Delta x = \Delta y = \Delta z = 5$ nm.

Fig. 4.1 Computation time

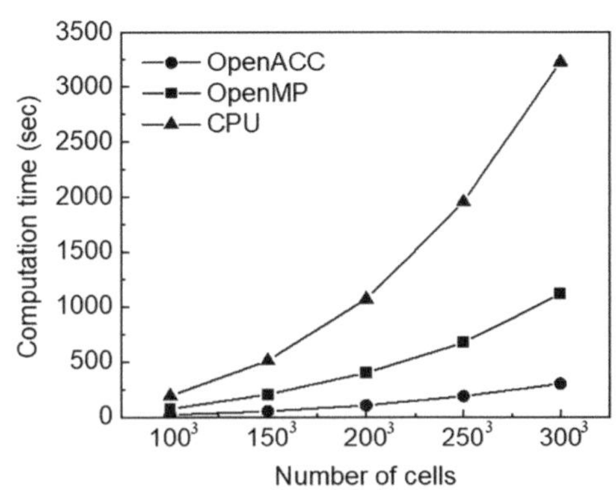

The calculations are performed as a function of the number of cells, using double-precision operations. Figure 4.1 shows the computation time of the GPU- and CPU-based FDTD methods, in which only single core is used for the CPU calculations. It is seen that the computation time for OpenACC is successfully reduced to less than 10% compared with the CPU calculations, demonstrating efficient calculations of practical 3-D plasmonic devices.

4.3 Slit-Type Filters

Using the GPU-FDTD method, we consider the bandpass characteristics of a slit-type filter in which the gap of the MIM waveguide is partially filled with metal (the 2-D structure was considered in [20]). The slit-type filter to be analyzed is shown in Fig. 4.2. The metal gap width at the input/output section is fixed at $w = 0.05$ μm. The metal thickness is set to $h = 0.05$ μm, which is widely used in 3-D plasmonic gap waveguides [2]. The length and width of the slit are set to $L_s = 0.3$ μm and $W_s = 0.05$ μm, respectively. The gap between the slit and the waveguide is $W_g = 0.02$ μm, and the metal wall between the input and output sections is $L_m = 0.05$ μm. The midpoints of L_s and L_m coincide in the z direction. The refractive index at the slit is n_S. The wavelength response at the output port is calculated using the Fourier transform along with the pulse excitation technique.

The transmission spectrum at the output port of the 3-D filter is shown in Fig. 4.3 for $n_S = 1$. For comparison, the results for the 2-D filter ($h = \infty$) are also

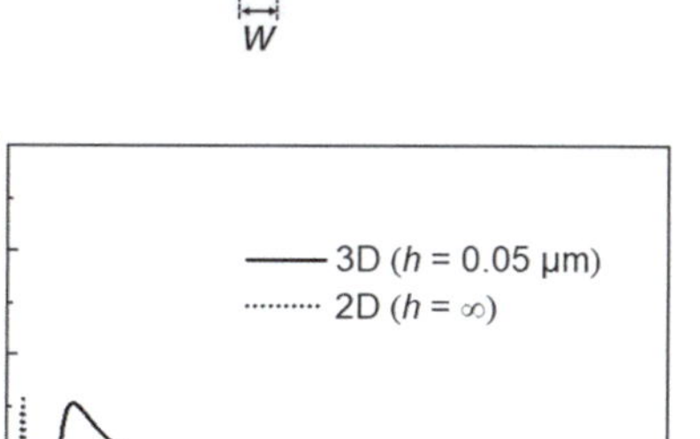

Fig. 4.2 Slit-type filter. Reproduced from [27], Copyright 2019, with permission from Elsevier

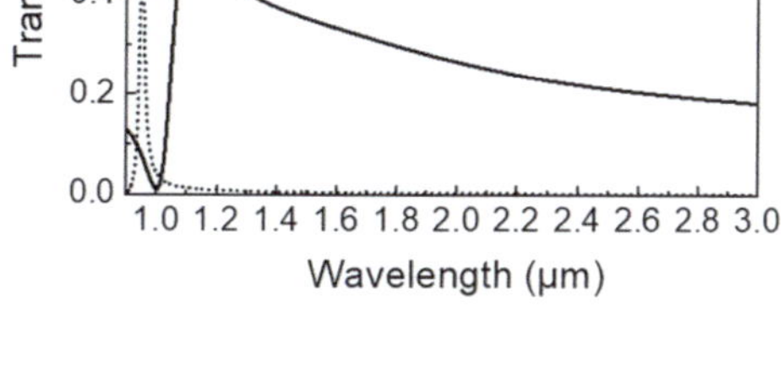

Fig. 4.3 Transmission spectra for $n_S = 1$. Reproduced from [27], Copyright 2019, with permission from Elsevier

shown. It can be seen that the 2-D filter exhibits a narrow transmission peak at 2% bandwidth. This is due to the fact that the electric field in the 2-D structure is completely confined to the waveguide and slit, and there is no field radiation. Unfortunately, the 3-D filter does not have sufficient bandpass characteristics, resulting in an undesirable transmission (94% bandwidth) on the long wavelength side. This undesirable transmittance is due to the coupling of the radiation field generated at the waveguide discontinuities to the output port.

To achieve a narrower bandwidth for the 3-D filter, it is necessary to increase the peak transmittance and suppress undesirable transmission on the long wavelength side. Here, the peak transmittance is first increased by filling the slit section with a dielectric material. A material with a Sellmeier coefficient that can be implemented by the TRC algorithm (Equation 8 in [34]) can be used, but here only materials with constant refractive index are used for simplicity. $n_S = 1.45$ (SiO_2), 2.60 (SiC), and 3.47 (Si) are chosen, and the transmission spectra for several n_S values are shown in Fig. 4.4. The transmission peak tends to increase as n_S is increased, but at $n_S = 3.47$, the transmission is slightly reduced due to high propagation loss. Therefore, $n_S = 2.60$, which has a maximum transmission of 81%, is selected. This high transmittance is due to the enhanced field confinement to the slit by the dielectric insertion, leading to a reduction in the radiated field [35].

The next step is to reduce the undesirable transmission at longer wavelengths. To do so, the radiated power coupled to the output port must be further reduced. In the case of a 2-D filter ($h = \infty$), there is no radiated field generated at the discontinuities as described above. Therefore, it can be inferred that the radiated field can be reduced by using a thick metal layer even in a 3-D structure, but a device as thin as possible is desirable. Figure 4.5 shows the transmission spectrum of a filter with a thicker metal thickness ($n_S = 2.60$), where the transmittance is normalized to the maximum value. It can be seen that as h is increased, the undesirable transmittance decreases on the long wavelength side. This is a natural result because the waveguide becomes closer to a two-dimensional structure. However, as h is increased, the peak transmittance decreases; for example, at $h = 0.4$ μm, the peak transmittance is less than 50%.

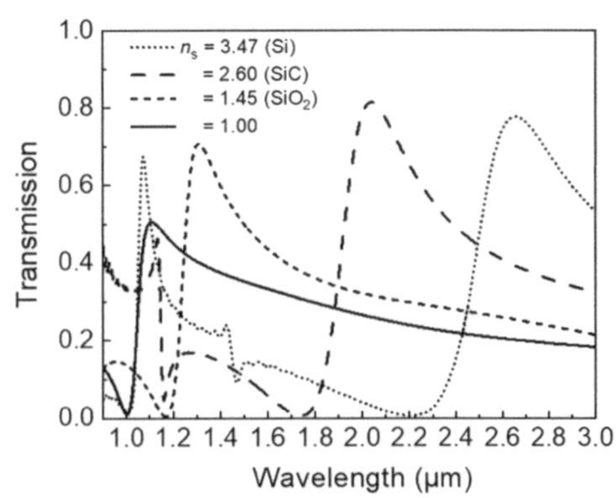

Fig. 4.4 Transmission spectra for several n_S values. Reproduced from [27], Copyright 2019, with permission from Elsevier

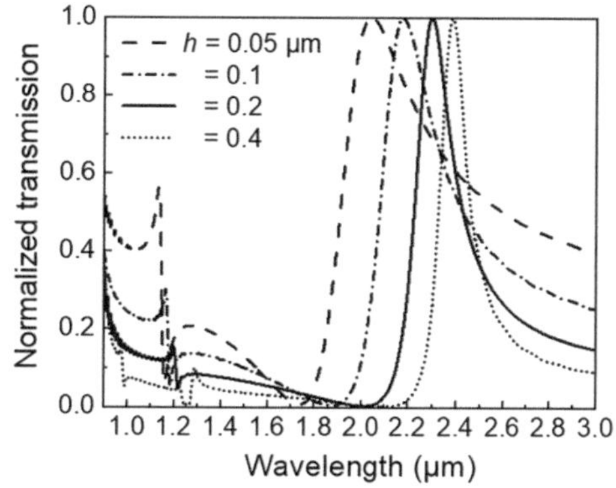

Fig. 4.5 Transmission spectra for several metal thicknesses. Reproduced from [27], Copyright 2019, with permission from Elsevier

Therefore, here we chose $h = 0.2\ \mu\text{m}$, which has a transmittance of more than 60%, and reduced the bandwidth from 94% in Fig. 4.3 to 9%.

Next, we adjust the slit length for transmittance at 1.31 and 1.55 μm. Figure 4.6 shows the transmittance for $L_S = 0.150$ and 0.185 μm. It can be seen that more than 60% output is successfully obtained for both wavelengths.

4.4 Wavelength Demultiplexer

We here consider a wavelength demultiplexer with two filters in parallel. Figure 4.7 shows an example of a demultiplexer designed to output a wavelength of 1.31 μm from port 1 and 1.55 μm from port 2. The structural parameters w, W_g and L_m are the same as in Fig. 4.2, and the slit length and metal thickness are $L_1 = 0.150\ \mu\text{m}$, $L_2 = 0.185\ \mu\text{m}$, and $h = 0.2\ \mu\text{m}$. The other parameters are $L_0 = 0.325\ \mu\text{m}$, $d_0 = 0.21\ \mu\text{m}$. Figure 4.8 shows the transmission spectra evaluated from the two output ports. It can be seen that the separation at 1.31 μm between the

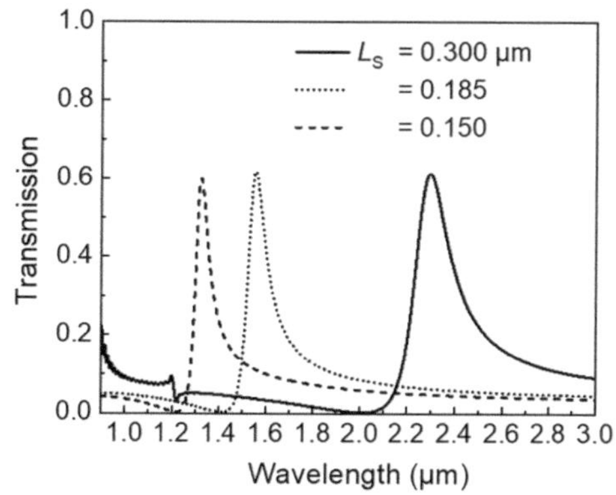

Fig. 4.6 Transmission spectra for several slit length ($h = 0.2\ \mu\text{m}$). Reproduced from [27], Copyright 2019, with permission from Elsevier

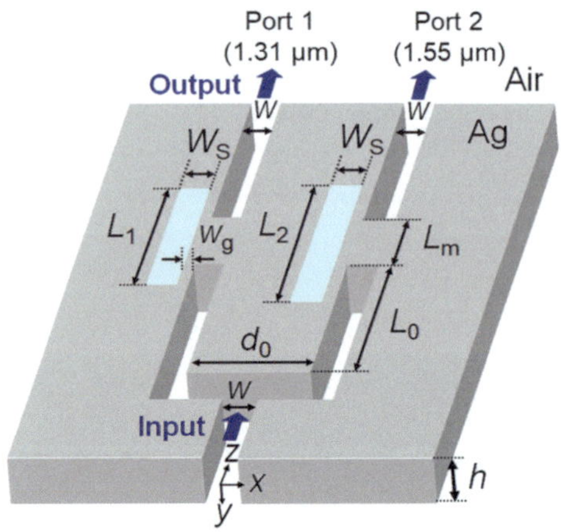

Fig. 4.7 Wavelength demultiplexer. Reproduced from [27], Copyright 2019, with permission from Elsevier

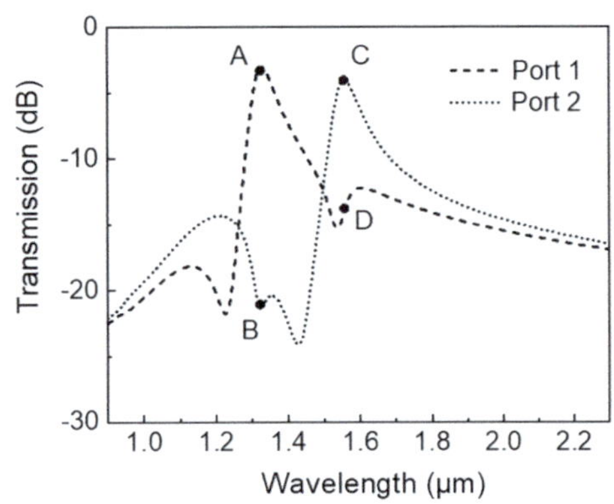

Fig. 4.8 Transmission spectra at the two output ports. Reproduced from [27], Copyright 2019, with permission from Elsevier

two ports is calculated to be 18.9 dB (A-B), while the separation at 1.55 μm is only 9.8 dB (C-D).

Note that in Fig. 4.8, the channel separation is not sufficient, less than 20 dB. To improve this separation, a new stub section that acts as an attenuation pole is introduced [12–14, 36, 37]. Here, we check the transmission characteristics of the stub-type bandstop filter with SiC also filling the stub, as shown in the inset of Fig. 4.9. Figure 4.9 shows the transmission with structural parameters of $W_{st} = 0.05$ μm, $L_{st} = 0.275$, 0.34 μm, and $h = 0.2$ μm. It can be seen that attenuation is obtained at 1.31 and 1.55 μm.

Wavelength multiplexers with stubs introduced in each waveguide are used at $W_{st1} = W_{st2} = 0.05$ μm, $L_{st1} = 0.34$, $L_{st2} = 0.275$ μm, and $L_d = 0.125$ μm as shown in Fig. 4.10. Figure 4.11 shows the transmission spectrum of the demultiplexer, which shows an improvement from 21.6 dB (A-E) at 1.31 μm to 40.1 dB (F-G) at 1.55 μm . This is a better separation than that of a 3-D MIM demultiplexer with ring resonators (about 3 dB at 1.31mm (Fig. 6 in [38])).The reflectance at the

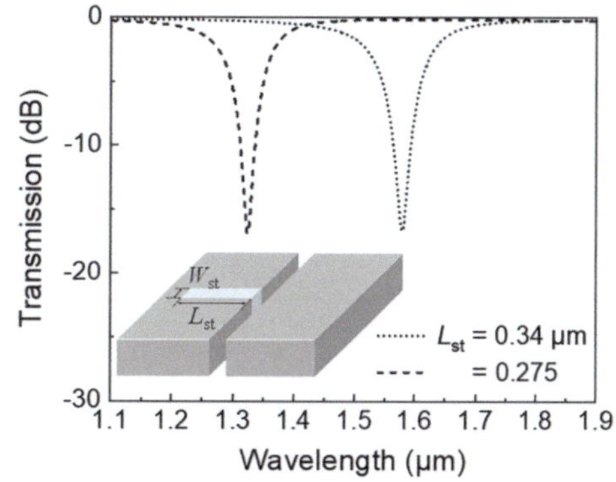

Fig. 4.9 Transmission spectra of a stub-type filter. Reproduced from [27], Copyright 2019, with permission from Elsevier

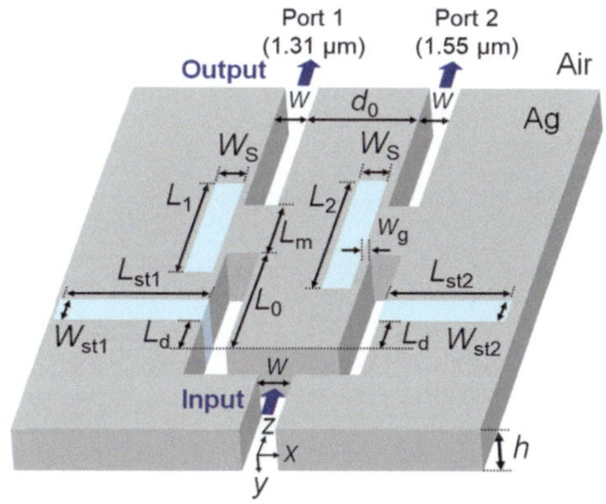

Fig. 4.10 Wavelength demultiplexer with stubs. Reproduced from [27], Copyright 2019, with permission from Elsevier

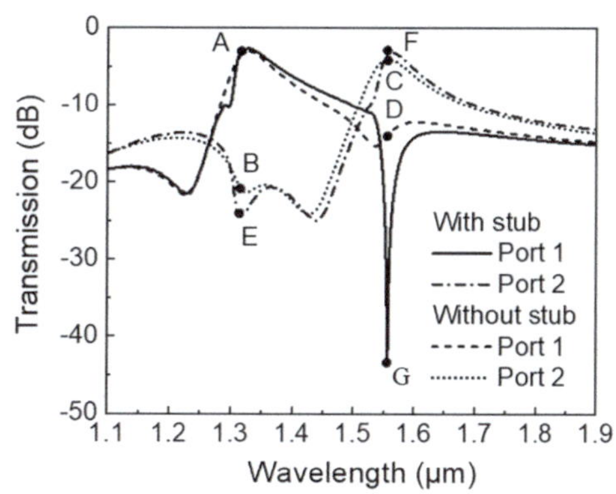

Fig. 4.11 Transmission spectra at the two output ports for the demultiplexer with stubs. Reproduced from [27], Copyright 2019, with permission from Elsevier

input port is shown in Fig. 4.12, and is reduced to less than −20 dB at both wavelengths. This result demonstrates the potential of the 3-D structure of the plasmon demultiplexer to achieve port separations of 20 dB or better. The field snapshots in the x-z plane at $y = 0$ are shown in Fig. 4.13a and b at 1.31 and 1.55 μm, respectively, clearly confirming the demultiplexing behavior.

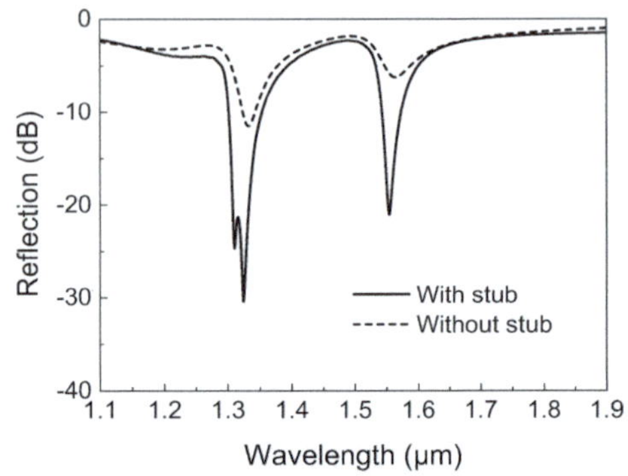

Fig. 4.12 Reflection spectra at the input port. Reproduced from [27], Copyright 2019, with permission from Elsevier

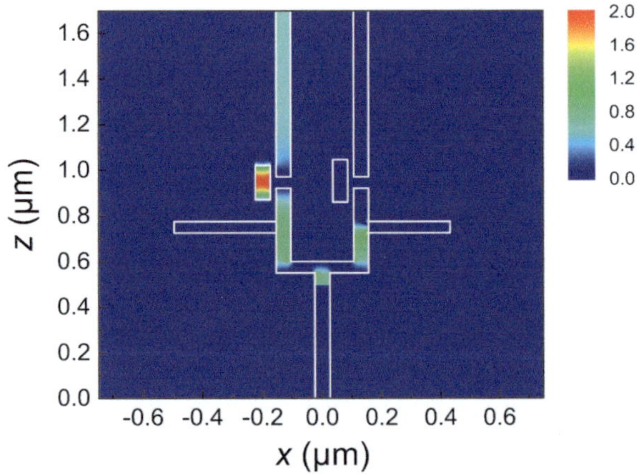

(a) 1.31 μm

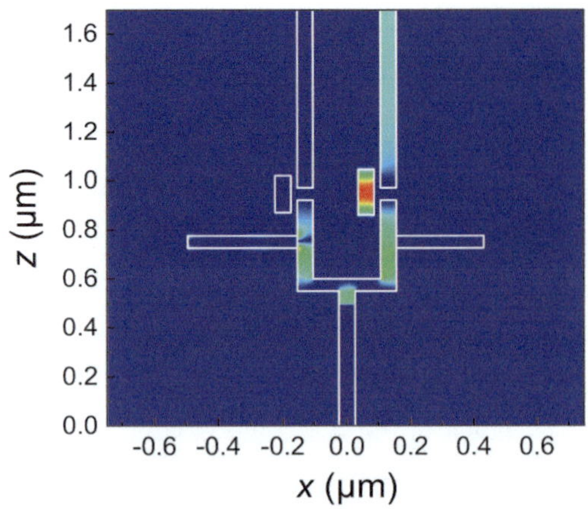

(b) 1.55μm

Fig. 4.13 Field distributions

Finally, we comment on the fabrication tolerance of this demultiplexer. The vacuum deposition method is widely used to form metal films. With this technique, the thickness of the metal can be controlled on the order of one nanometer. Therefore, the demultiplexer is expected to be fabricated as designed. Nevertheless, it is necessary to know the degree to which the characteristics deteriorate when the thickness is varied. Therefore, we calculated the demultiplexer for a metal thickness of 0.205 μm with a shift of 2.5% from the designed value. The results show that the peak frequency shift is relatively small, 0.133% at 1.31 μm and 0.156% at 1.55 μm, indicating high fabrication tolerance.

4.5 Concluding Remarks

We have studied a wavelength demultiplexer based on slit-type filters composed of 3-D MIM plasmonic waveguides using a GPU-accelerated TRC-FDTD method. First, to obtain a narrow bandwidth of the 3-D slit-type filter, we insert a dielectric material into the slit and choose a metal thickness of 0.2 μm. The slit length is adjusted to obtain transmittance at 1.31 and 1.55 μm. Then, the two filters are placed in parallel to form a demultiplexer. It is shown that a separation of 18.9 dB is obtained between the two output ports at 1.31 μm, while it is 9.8 dB at 1.55 μm. To improve the separation characteristics, stubs forming an attenuation pole are inserted into the waveguides. It is found that the separation is successfully improved to more than 20 dB at both wavelengths, with a reduced reflection power of less than -20 dB.

References

1. K. Tanaka, M. Tanaka, Simulations of nanometric optical circuits based on surface plasmon polariton gap waveguide. Appl. Phys. Lett. **82**, 1158–1160 (2003)
2. G. Veronis, S. Fan, Modes of subwavelength plasmonic slot waveguides. J. Lightwave Technol. **25**, 2511–2521 (2007)
3. A. Hosseini, Y. Massoud, Nanoscale surface plasmon based resonator using rectangular geometry. Appl. Phys. Lett. **90**, 181102-1–3 (2007)
4. T.-B. Wang, X.-W. Wen, C.-P. Yin, H.-Z. Wang, The transmission characteristics of surface plasmon polaritons in ring resonator. Opt. Express **17**, 24096–24101 (2009)
5. G. Wang, H. Lu, X. Liu, D. Mao, L. Duan, Tunable multi-channel wavelength demultiplexer based on MIM plasmonic nanodisk resonators at telecommunication regime. Opt. Express **19**, 3513–3518 (2011)
6. M.R. Rakhshani, A. Tavousi, M.A. Mansouri-Birjandi, Design of a plasmonic sensor based on a square array of nanorods and two slot cavities with a high figure of merit for glucose concentration monitoring. Appl. Opt. **57**, 7798–7804 (2018)
7. A. Hosseini, Y. Massoud, A low-loss metal-insulator-metal plasmonic bragg reflector. Opt. Express **14**, 11318–11323 (2006)
8. Z. Han, E. Forsberg, S. He, Surface plasmon Bragg gratings formed in metal-insulator-metal waveguides. IEEE Photonics Technol. Lett. **19**, 91–93 (2007)
9. J.-Q. Liu, L.-L. Wang, M.-D. He, W.-Q. Huang, D. Wang, B.S. Zou, S. Wen, A wide bandgap plasmonic Bragg reflector. Opt. Express **16**, 4888–4893 (2008)

10. J. Park, H. Kim, B. Lee, High order plasmonic Bragg reflection in the metal-insulator-metal waveguide Bragg grating. Opt. Express **16**, 413–425 (2008)
11. J. Shibayama, A. Nomura, R. Ando, J. Yamauchi, H. Nakano, A frequency-dependent LOD-FDTD method and its application to the analyses of plasmonic waveguide devices. IEEE J. Quantum Electron. **46**, 40–49 (2010)
12. Y. Matsuzaki, T. Okamoto, M. Haraguchi, M. Fukui, M. Nakagaki, Characteristics of gap plasmon waveguide with stub structures. Opt. Express **16**, 16314–16325 (2008)
13. A. Pannipitiya, I.D. Rukhlenko, M. Premaratne, H.T. Hattori, G.P. Agrawal, Improved transmission model for metal-dielectric-metal plasmonic waveguides with stub structure. Opt. Express **18**, 6191–6204 (2010)
14. S.R. Mirnaziry, A. Setayesh, M.S. Abrishamian, Design and analysis of plasmonic filters based on stubs. J. Opt. Soc. Am. B **28**, 1300–1307 (2011)
15. J. Tao, X.G. Huang, X. Lin, Q. Zhang, X. Jin, A narrow-band subwavelength plasmonic waveguide filter with asymmetrical multipleteeth-shaped structure. Opt. Express **17**, 13989–13994 (2009)
16. X.-S. Lin, X.-G. Huang, Tooth-shaped plasmonic waveguide filters with nanometeric sizes. Opt. Lett. **33**, 2874–2876 (2008)
17. J. Yu, Y. Ohtera, H. Yamada, Scattering-parameter model analysis of side-coupled plasmonic Fabry-Perot waveguide filters. Appl. Phys. A **124**, 1–7 (2018)
18. H. Li, R.-Z. Jiao, Plasmonic band-stop filters based on tooth structure. Opt. Commun. **439**, 201–205 (2019)
19. A. Noual, A. Akjouj, Y. Pennec, J.-N. Gillet, B. Djafari-Rouhani, Modeling of two-dimensional nanoscale Y-bent plasmonic waveguides with cavities for demultiplexing of the telecommunication wavelengths. N. J. Phys. **11**, 103020-1–19 (2009)
20. X. Mei, X. Huang, J. Tao, J. Zhu, Y. Zhu, X. Jin, A wavelength demultiplexing structure based on plasmonic MDM side-coupled cavities. J. Opt. Soc. Am. B **27**, 2707–2713 (2010)
21. J. Chen, Z. Li, M. Lei, X. Fu, J. Xiao, Q. Gong, Plasmonic Y-splitters of high wavelength resolution based on strongly coupled-resonator effects. Plasmonics **7**, 441–445 (2011)
22. Y. Guo, L. Yan, W. Pan, B. Luo, K. Wen, Z. Guo, H. Li, X. Luo, A plasmonic splitter based on slot cavity. Opt. Express **19**, 13831–13838 (2011)
23. F. Hu, H. Yi, Z. Zhou, Wavelength demultiplexing structure based on arrayed plasmonic slot cavities. Opt. Lett. **36**, 1500–1502 (2011)
24. D. Xianga, W. Li, MIM plasmonic waveguide splitter with tooth-shaped structures. J. Modern Opt. **61**, 222–226 (2014)
25. M. Reza Pav, N. Granpayeh, S. Poorya Hosseini, A. Rahimzadegan, Ultracompact double tunable two-channel plasmonic filter and 4-channel multi/demultiplexer design based on aperture-coupled plasmonic slot cavity. Opt. Commun. **437**, 285–289 (2019)
26. M. Bahramipanaha, M.S. Abrishamiana, S.A. Mirtaheria, J.-M. Liu, Ultracompact plasmonic loop-stub notch filter and sensor. Sens. Actuators B **194**, 311–318 (2014)
27. J. Shibayama, H. Kawai, J. Yamauchi, H. Nakano, Analysis of a 3D MIM waveguide-based plasmonic demultiplexer using the TRC-FDTD method. Opt. Comm. **452**, 360–365 (2019)
28. J. Shibayama, R. Ando, J. Yamauchi, H. Nakano, Frequency-dependent 3-D LOD-FDTD method for the analysis of plasmonic devices. IEEE Photonics Technol. Lett. **23**, 1070–1072 (2011)
29. J. Shibayama, T. Hirano, J. Yamauchi, H. Nakano, Efficient implementation of frequency-dependent 3-D LOD-FDTD method using fundamental scheme. Electron. Lett. **48**, 774–775 (2012)
30. P. Sypek, A. Dziekonski, M. Mrozowski, How to render FDTD computations more effective using a graphics accelerator. IEEE Trans. Magnetics **45**, 1324–1327 (2009)
31. D. De Donno, A. Esposito, L. Tarricone, L. Catarinucci, Introduction to GPU computing and CUDA programming: a case study on FDTD. IEEE Ant. Propag. Mag. **52**, 116–122 (2010)
32. S. Wienke, P. Springer, C. Terboven, D. an Mey, OpenACC–First experiences with real-world applications. Europ. Con. Parallel Proc. **7484**, 859–870 (2012)

33. D.F. Kelley, R.J. Luebbers, Piecewise linear recursive convolution for dispersive media using FDTD. IEEE Trans. Antennas Propag. **44**, 792–797 (1996)
34. R. Siushansian, J. LoVetri, Efficient evaluation of convolution integrals arising in fdtd formulations of electromagnetic dispersive media. J. Electromagnet. Wave **11**, 101–117 (1997)
35. J. Shibayama, Y. Wada, J. Yamauchi, H. Nakano, Analysis of two- and three-dimensional plasmonic waveguide band-pass filters using the TRC-FDTD method. IEICE Trans. Electron. **E99-C**, 817–819 (2016)
36. J. Shibayama, H. Kawai, J. Yamauchi, H. Nakano, Numerical investigation of a three-dimensional stub-type plasmonic filter, in *16th International Conference on Numerical Simulation of Optoelectronic Devices, MP*, vol. 20, pp. 67–68 (2016)
37. S. Naghizadeh, ŞE. Kocabaş, Guidelines for designing 2D and 3D plasmonic stub resonators. J. Opt. Soc. Am. B **34**, 207–217 (2017)
38. T. Nurmohammadia, K. Abbasianb, R. Yadipoura, A proposal for a demultiplexer based on plasmonic metal-insulator-metal waveguide-coupled ring resonator operating in near-infrared spectrum. Optik **142**, 550–556 (2017)

Chapter 5
Surface Plasmon Resonance Waveguide Sensor at Terahertz Frequencies

Abstract This chapter presents the analysis of a surface plasmon resonance (SPR) waveguide sensor with a thin InSb layer in the sensing section in the terahertz region. First, the eigenmode analysis of the sensing section is performed using water as the analyte. Then, the eigenmodes are calculated using the Yee-mesh based imaginary distance beam propagation method. As a result, the eigenmodes of the surface plasmon polariton exist in the sensing region, and the maximum SPR is expected to be around 1.28 THz at 300 K. Next, we focus on the treatment of the InSb-water interface and investigate the performance of the sensor using the frequency-dependent FDTD method based on the simple trapezoidal recursive convolution technique. As expected, the SPR absorption is found to be maximum at 1.26 THz. Furthermore, the SPR response is investigated and found to be capable of detecting temperature variations from 280 to 320 K. As an application, the possibility of detecting mixtures of organic solvents and water is investigated. The sensor with an analyte container is also investigated. A steep frequency response is achieved with a low height incidence waveguide. Finally, a practical three-dimensional sensor is also investigated.

5.1 Introduction

In the optical frequency band, surface plasmon resonance (SPR) sensors have been extensively studied and are already commercially available [1–4]. SPR sensors utilize surface plasmon polariton (SPP) waves excited along the metal/dielectric interface, the properties of which are highly sensitive to characteristics of the analyte.

Conventional SPR sensors have widely used the Kretschmann configuration [5], a simple structure consisting of a prism coated with a thin metal film. However, the Kretschmann configuration is not suitable for monolithic integration into optical circuits due to the bulk structure of the prism. Recently, sensors based on SPR waveguides have been investigated [6–10] and monolithic integration is now possible using optical device fabrication techniques. We have theoretically investigated several SPR waveguide sensors and improved their performance [11–14].

Optical frequency SPR sensors can detect, for example, protein-DNA interactions [15] and reactions of antigen-antibody complexes [16]. In terahertz waves,

J. Shibayama, *FDTD Analysis of Plasmonic Devices*, Springer Series in Optical Sciences 254, https://doi.org/10.1007/978-981-96-8130-3_5

on the other hand, hydrogen vibration modes exist in the vicinity of 1 THz. In addition, terahertz waves are strongly absorbed by pure water, but not by aqueous solutions. Therefore, the use of terahertz waves has attracted attention in the field of biochemistry.

At THz frequencies, SPP waves are not supported along the metal/dielectric interface. This is because the THz frequency is much lower than the plasma frequency of the metal, so the penetration of electromagnetic waves is negligible. It should be noted that semiconductor materials such as indium antimonide (InSb) exhibit plasma frequencies in the THz band [17–20]. This makes it possible to propagate THz-band SPP waves along the InSb/dielectric interface. Several groups have used these SPP waves to study temperature-dependent THz functional devices [21–27]. However, little attention has been paid to the study of SPR waveguide sensors in the THz region.

Here, we propose a THz SPR waveguide sensor that employs a thin InSb layer in the sensing part [28]. After reviewing the frequency and temperature dependent permittivities of InSb and water, we perform the eigenmode analysis for the sensing section. Numerical results show that SPP eigenmodes exist in the sensing section with water. The overlap integral calculations between the SPP and input eigenmodes indicate that the SPR absorption peak is expected to be around 1.28 THz at 300 K. Then, the sensor is analyzed using the frequency-dependent FDTD method, paying attention to the treatment of the interface between InSb and water. In order to obtain higher numerical accuracy for a small field amplitude, the interface is expressed by the Drude model, which has an average permittivity of InSb and water. As expected from the eigenmode analysis, the maximum SPR absorption is found to occur at 1.26 THz. Furthermore, the SPR response is investigated for a temperature variation from 280 to 320 K. The application of the present SPR sensor to the detection of an aqueous solution of an organic solvent is also considered. In particular, an acetonitrile-water solution is considered and the ability to detect its volume fractions is demonstrated.

Moreover, an SPR sensor with an analyte container is examined, with respect to the coupling of the incident field to the sensor section. It is shown that a material with a lower refractive index is desirable for the analyte container. As a result, COP is chosen as the container. It is also found that the lower-height of the incidence waveguide compared to that in the sensing section contributes to a steep frequency response of the output power.

Finally, the analysis of a practical 3-D sensor with an analyte container is performed. The size of the container is found by the full 3-D FDTD analysis. The characteristics of the 3-D sensor are found to be almost the same as the 2-D sensor. Note that since the frequency response of the 3-D sensor is different from the 2-D counterpart, we need to perform the 3-D analysis for a practical design of the 3-D sensor structures.

5.2 Permittivities of InSb and Water

The dielectric constants of InSb and water exhibit frequency and temperature dependence. Dispersion models for these two materials are outlined.

The permittivity of InSb is often expressed by the Drude model as [19–27]

$$\varepsilon_{\mathrm{I}}(\omega) = \varepsilon_{\infty} + \frac{\omega_{\mathrm{D}}^2}{j\omega\nu_{\mathrm{D}} - \omega^2} \tag{5.1}$$

where $\varepsilon_{\infty}(=15.7)$ is the high-frequency permittivity, ω is the angular frequency, ω_{D} is the plasma frequency, and ν_{D} is the electron collision rate. The latter two are defined as

$$\omega_{\mathrm{D}} = \sqrt{\frac{N_d e^2}{\varepsilon_0 m^*}} \tag{5.2}$$

$$\nu_{\mathrm{D}} = \frac{e}{m^* \mu} \tag{5.3}$$

where N_d is the free carrier concentration, e is the elementary charge, ε_0 is the vacuum permittivity, m^* is the effective mass of charge carrier, and μ is the mobility of charge carrier. The electron effective mass of InSb is $0.014m_0$, where m_0 is the electron mass. For InSb, N_d and μ can be obtained as [26]

$$N_d = 6 \times 10^{14} T^{3/2} e^{-0.29/2kT} \ (\mathrm{cm}^{-3}) \tag{5.4}$$

$$\mu = 7.7 \times 10^4 \left(\frac{T}{300}\right)^{-1.66} \ (\mathrm{cm}^2/\mathrm{Vs}) \tag{5.5}$$

where T and k are the temperature and Boltzmann constant, respectively.

Figure 5.1 shows ω_{D} and ν_{D} as a function of temperature obtained from (5.2) and (5.3) [26], in which the symbols denote the data available from [19]. A close correspondence can be seen to exist between the results from [26] and those from [19]. To investigate the characteristics of the sensor for any given temperature, (5.2) and (5.3) are used here.

The relative permittivity of water is expressed by the following double Debye model [29]:

$$\varepsilon_{\mathrm{w}}(\omega) = \varepsilon_2 + \frac{\varepsilon_0 - \varepsilon_1}{1 + \frac{j\omega}{2\pi\gamma_1}} + \frac{\varepsilon_1 - \varepsilon_2}{1 + \frac{j\omega}{2\pi\gamma_2}} \tag{5.6}$$

where

$$\varepsilon_0 = 77.66 - 103.3\theta$$
$$\varepsilon_1 = 0.0671\varepsilon_0$$

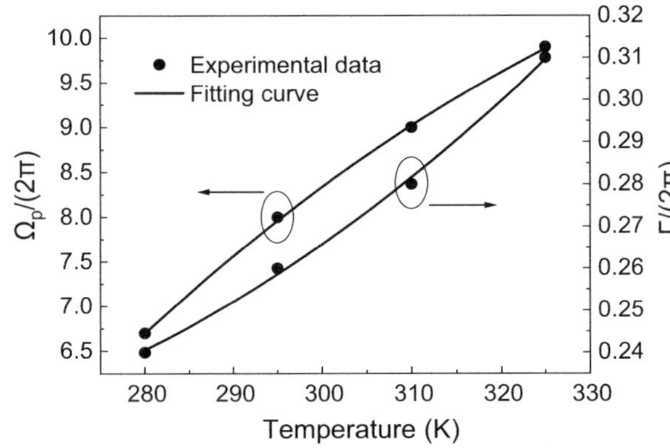

Fig. 5.1 ω_D and ν_D adopted for the Drude model of InSb. ©[2016] IEEE. Reproduced, with permission, from [28]

$$\varepsilon_2 = 3.52 + 7.52\theta$$
$$\gamma_1 = 20.20 + 146.4\theta + 316\theta^2$$
$$\gamma_2 = 39.8\gamma_1$$
$$\theta = 1 - \frac{300}{T}$$

in which ε_0 is the static dielectric constant of water, ε_1 and ε_2 are respectively the first and second high-frequency constants, γ_1 and γ_2 are respectively the primary and secondary relaxation frequencies, and θ is the modified relative inverse variable. The permittivity from (6) follows the experimental data quite well [29]. The above parameters of InSb and water are incorporated into the frequency-dependent FDTD method.

5.3 Eigenmode Analysis of the Sensing Section

The sensing performance depends mainly on the propagation characteristics of the SPP waves. Therefore, it is important to know the SPP characteristics in the sensing section at THz frequencies. Here, we perform the eigenmode analysis in the sensing section using the imaginary-distance finite-difference beam-propagation method (BPM) [30, 31]. The imaginary-distance method efficiently provides eigenmodes by changing the coordinate in the propagation direction of the BPM to $j\tau$.

The 2-D sensor configuration is shown in Fig. 5.2, where the thin InSb layer is placed on the Si/SiO_2 waveguide and water is chosen as the analyte. In this paper, Si and SiO_2 are assumed to be temperature independent and lossless materials, and their relative permittivities are chosen to be 3.4^2 [32] and 1.95^2 [33], respectively. The configuration parameters are $d = 50$ μm, $t_I = 5$ μm and $l_a = 400$ μm (preliminary calculations show that a thickness of $t_I = 5$ μm exhibits a strong SPR resonance over a wide frequency range). The temperature is set to $T = 300$ K.

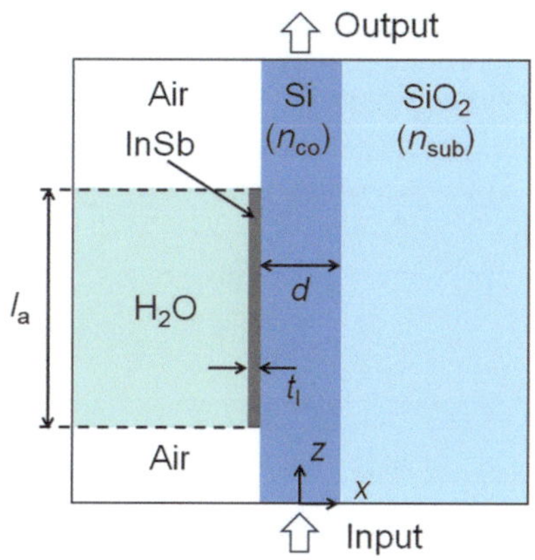

Fig. 5.2 Configuration of an SPR waveguide sensor. ©[2016] IEEE. Reproduced, with permission, from [28]

As a result of the analysis, we observed three SPP modes in the sensor section. Among them, the lowest mode has a large propagation loss and a negligible coupling from the input eigenmode. Therefore, we discuss only the two higher eigenmodes, where one with a larger effective index is called the first mode and the other is called the second mode.

The two SPP eigenmode fields (H_y and E_x components) at 1.28 THz are shown in Fig. 5.3, with the field amplitude normalized to its maximum value. It can be seen that the field amplitude is significantly enhanced around the thin InSb layer. This results in a large evanescent field in the analyte, leading to the ability to detect the analyte property. The effective index and propagation loss versus frequency are shown in Fig. 5.4. As can be seen, the effective indices for both modes increase with frequency, while the propagation losses are almost the same at $\simeq$1.28 THz.

To estimate how much the input eigenmode is coupled to each of the two SPP modes, we calculate the overlap integral between them. The overlap integral is defined as

$$O.I. = \frac{|\int \phi_{\mathrm{SPP}} \cdot \phi_{\mathrm{Input}}^{*} dx|^2}{\int |\phi_{\mathrm{SPP}}|^2 dx \int |\phi_{\mathrm{Input}}|^2 dx} \tag{5.7}$$

where ϕ_{SPP} is the SPP mode field in the sensing section and ϕ_{Input} is the input eigenmode field. Although either the electric or magnetic mode field can be used for this calculation, the magnetic field is used in this paper.

Figure 5.5 shows the overlap integral, which also includes the sum of the two results. It can be seen that the total value reaches 0.9 at frequencies higher than 1.2 THz and becomes 0.98 at 1.28 THz, where the propagation losses of the two modes are almost equal. Note for the SPR sensor in the optical regime that the SPR absorption becomes maximum around the frequency where the propagation losses are almost equal for the two SPP modes [12]. For the present sensor, an SPR absorption peak is also expected to exist around 1.28 THz, where most of the input eigenmode is coupled to the two SPP modes and the two propagation losses are almost equal.

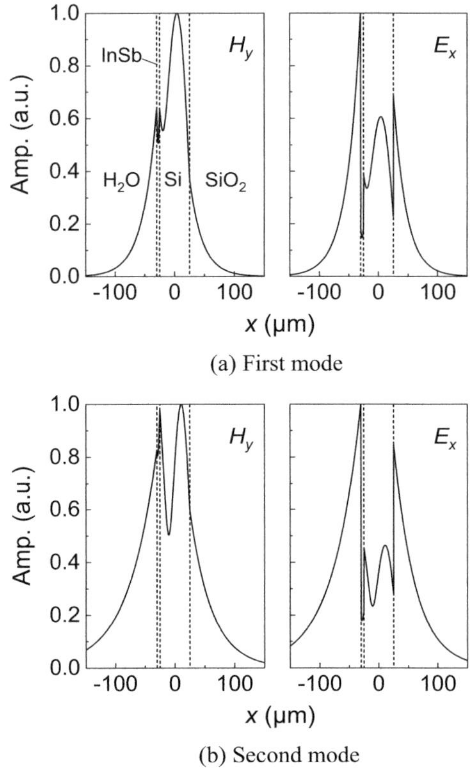

Fig. 5.3 Field distributions. ©[2016] IEEE. Reproduced, with permission, from [28]

5.4 Analysis of the 2-D SPR Waveguide Sensor

Since the SPP characteristics are clear from the eigenmode analysis, the SPR waveguide sensor is next analyzed using the FDTD method. For the FDTD analysis of dispersive media, the frequency-dependent version is required. In this paper, we use the frequency-dependent FDTD method based on the simple trapezoidal recursive convolution technique discussed in Chap. 2.

Note that for the present sensor, the Drude and Debye models form a boundary in the sensor section. Although the interface between two different dispersive materials should be carefully implemented for frequency-dependent FDTD analysis, its treatment has been little discussed. In the following, we discuss how to treat the interface between InSb and water.

For plasmonic devices, a dispersive material (e.g. metal at optical frequencies) is usually surrounded by a non-dispersive material (normal dielectric). In this case, we have two techniques to treat its interface. One is to calculate the electric field at the boundary with the dispersion model and that with the nondispersion model, followed by averaging these electric fields [34]. The other is to use a dispersion model with an averaged permittivity of the dispersion and nondispersion models [35, 36]. Note that

Fig. 5.4 Results of eigenmode analysis. ©[2016] IEEE. Reproduced, with permission, from [28]

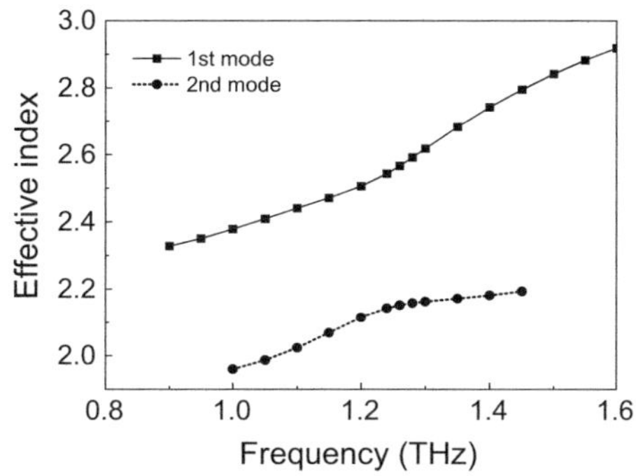

(a) Effective index

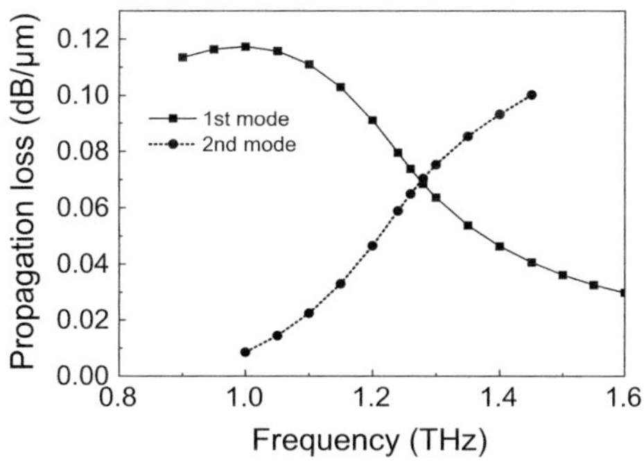

(b) Propagation loss

Fig. 5.5 Overlap integral. ©[2016] IEEE. Reproduced, with permission, from [28]

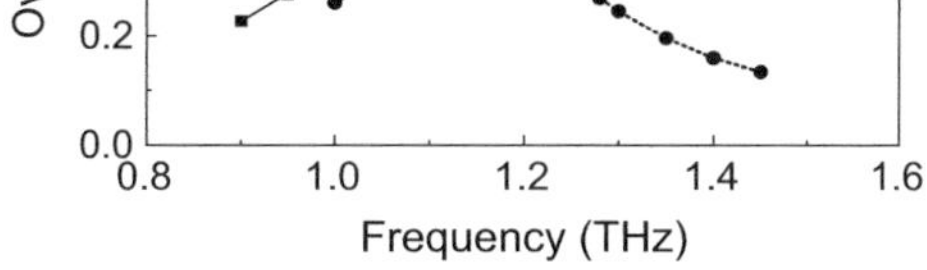

these two techniques have not been fully investigated for the case where two different dispersion models form a boundary. Therefore, here we investigate the performance of these two techniques for the InSb-water boundary. For convenience, the former and latter techniques will be referred to as Method 1 and Method 2, respectively.

Following Method 1, we average the electric field obtained from the Drude model (5.1) for InSb and the Debye model (5.6) for water at their interface. For Method 2, we assume that the permittivity at the interface is expressed by the Drude model with an averaged permittivity. We calculate the complex permittivities of InSb and water from (5.1) and (5.6) at a specific frequency to be investigated and then average these permittivities. By solving the simultaneous equations of the assumed Drude model and the averaged permittivities, we obtain the plasma frequency and the electron collision rate of the assumed Drude model with the averaged permittivity.

To examine the performance of the two methods, we compute the overlap integral between the eigenmode and the propagating field in the sensing section at 300 K. If the field is computed accurately, the eigenmode field does not deform with the overlap integral remaining constant (unity). Figure 5.6 shows the overlap integral versus propagation distance using the lowest eigenmode of the sensing section at 1 THz as the input field. It can be seen that the overlap integral of Method 1 deviates from unity around 200 μm. Since the boundary condition depends on the field amplitude, the numerical accuracy of method 1 cannot be maintained for a small field amplitude. For Method 2, on the other hand, the overlap integral maintains unity up to 450 μm. This is because the averaged permittivity does not depend on the field amplitude, leading to accurate results for a longer propagation distance.

Next, we compute the normalized power remaining in the computational domain to see to what extent numerical accuracy is maintained at small power. Figure 5.7 plots the normalized power versus propagation distance. From Figs. 5.6 and 6.7, Methods 1 and 2 maintain their accuracy with a normalized power of $\simeq -25$ and -70 dB, respectively.

Although Method 1 does not maintain its accuracy at small field amplitudes, it is relatively easy to implement, and its application to frequency response calculations is straightforward. While Method 2 provides higher numerical accuracy even at a small field amplitude, the derivation of the parameters requires solving simultaneous equations. In this paper, we use Method 2 to handle a small field amplitude, especially at the SPR frequency.

We investigate the SPR response of the sensor by launching the eigenmode of the input waveguide and evaluating the output power (guided mode power).

Fig. 5.6 Overlap integral

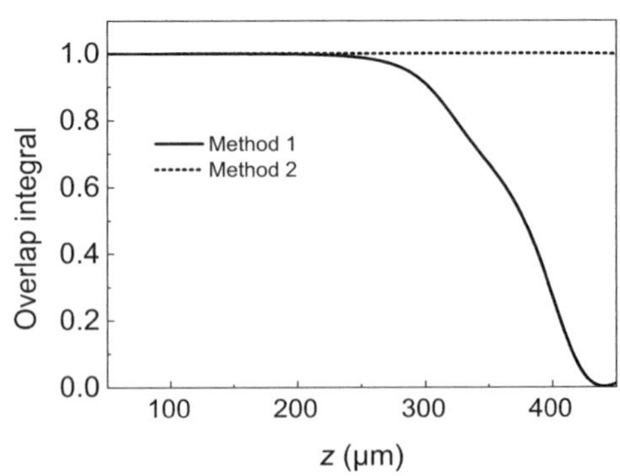

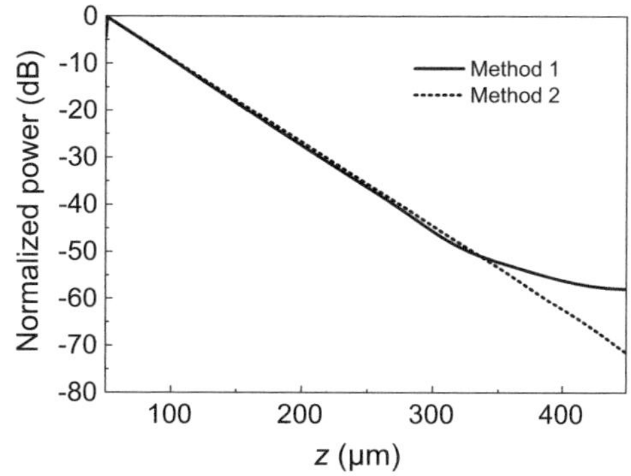

Fig. 5.7 Normalized power. ©[2016] IEEE. Reproduced, with permission, from [28]

Figure 5.8 shows the SPR response at $T = 300$ K. As expected from the eigenmode analysis, the maximum SPR absorption is found to occur at 1.26 THz. This frequency is close to the frequency at which the propagation losses of the two SPP modes are equal, as shown in Fig. 5.4b. Therefore, eigenmode analysis is useful for predicting the SPR frequency of a waveguide sensor.

Figure 5.9 shows the field distributions for the resonant peak at 1.26 THz and the off-resonant peak at 1.5 THz. It is observed that the input THz wave at 1.26 THz rapidly loses its power due to the strong SPR compared to the case at 1.5 THz. Here, we check the amplitude of the SPP in the sensing section. Figure 5.10 shows the propagating fields (H_y components) at $z = 200$ μm, where the field is normalized to its maximum value in the core. It can be seen that the SPP amplitude at 1.26 THz excited around the InSb layer is higher than that at 1.5 THz, resulting in a large propagation loss.

Since we have confirmed that the sensor has an SPR response, we next investigate the temperature dependence of the sensor. The SPR response is shown in Fig. 5.11, where the temperature ranges from 280 to 320 K. As can be seen, the peak frequency becomes higher as the temperature increases. Note that the real part of the InSb relative permittivity varies significantly from -16.3 to -35.3 as the temperature increases.

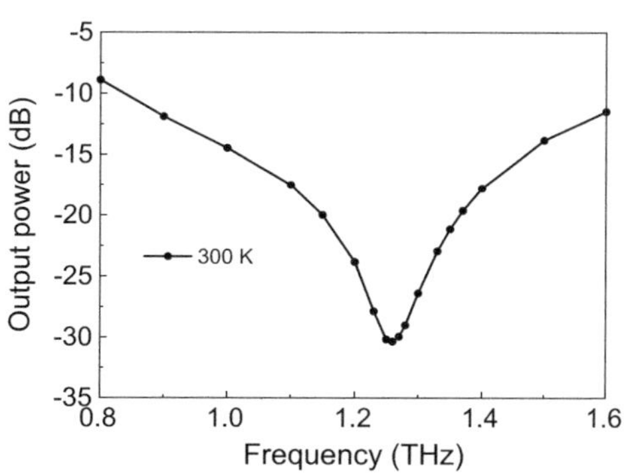

Fig. 5.8 Output power at $T = 300$ K. ©[2016] IEEE. Reproduced, with permission, from [28]

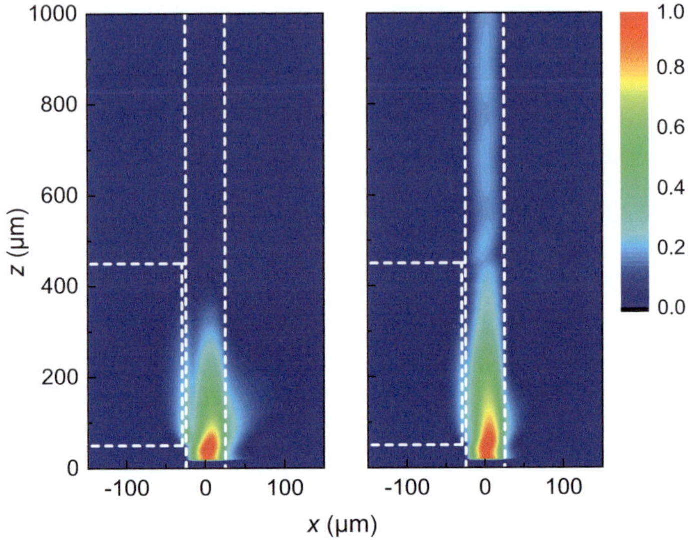

Fig. 5.9 Field distributions at $T = 300$ K. H_y components are displayed. ©[2016] IEEE. Reproduced, with permission, from [28]

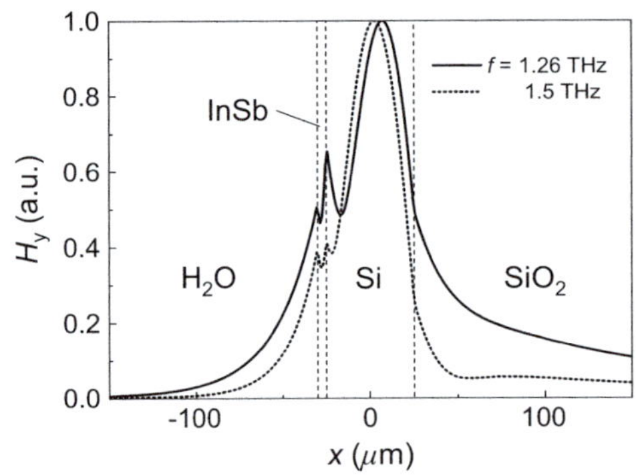

Fig. 5.10 Propagating fields at $z = 200$ μm. ©[2016] IEEE. Reproduced, with permission, from [28]

As a result, the shift in the peak frequency is mainly due to a change in the resonance condition resulting from the change in the InSb permittivity.

Figure 5.12 shows the peak frequency as a function of temperature. It can be seen that the peak frequency varies almost linearly with temperature. From this result, the sensitivity is estimated to be about 7.5×10^{-3} THz/K, which is higher than that reported in [27] (1.425×10^{-3} THz/K). This difference is probably due to the fact that water, with its temperature-dependent permittivity (5.6), increases the sensitivity of the present sensor. As a result, the temperature determines the SPR peak frequency, showing a possibility to act as a temperature sensor.

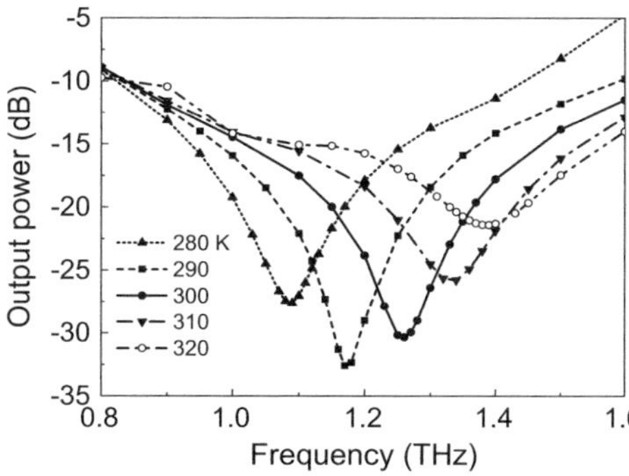

Fig. 5.11 SPR response for various temperatures. ©[2016] IEEE. Reproduced, with permission, from [28]

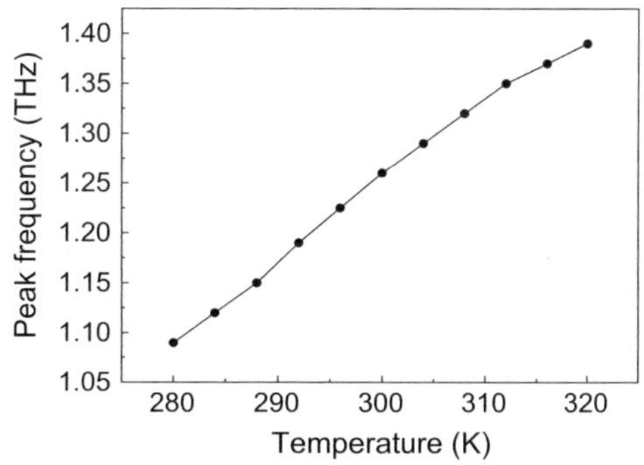

Fig. 5.12 Peak frequency versus temperature. ©[2016] IEEE. Reproduced, with permission, from [28]

Note that a structure to contain water has not been considered in this work. A candidate for this is the use of an analyte container that is a concave section made in the upper cladding [37]. We will apply the analyte container to the sensor later.

5.5 Detection of Mixture of Organic Solvent and Water

The application of detecting a mixture of an organic solvent and water using the present SPR waveguide sensor is considered. We consider an acetonitrile-water solution, which has received continuous attention for decades due to a wide variety of applications [38–41].

The permittivity of acetonitrile-water mixtures is expressed by the double Debye model, the parameters of which are given in [39]. Figure 5.13 shows the SPR response at $T = 293$ K for several volume fractions denoted by a_{ACN}. It can be seen that as a_{ACN} is increased, the output power also increases (the real part of the relative permittivity of water is reduced from 4.00 for $a_{\text{ACN}} = 0$ to 2.5 for $a_{\text{ACN}} = 1.0$, while the imaginary part is almost constant). This variation in output power is related to the strength of the SPR. That is, the SPR is strong for $a_{\text{ACN}} = 0$, resulting in the reduced output power. For $a_{\text{ACN}} = 1.0$, the SPR becomes weak, reducing the absorption of THz waves with

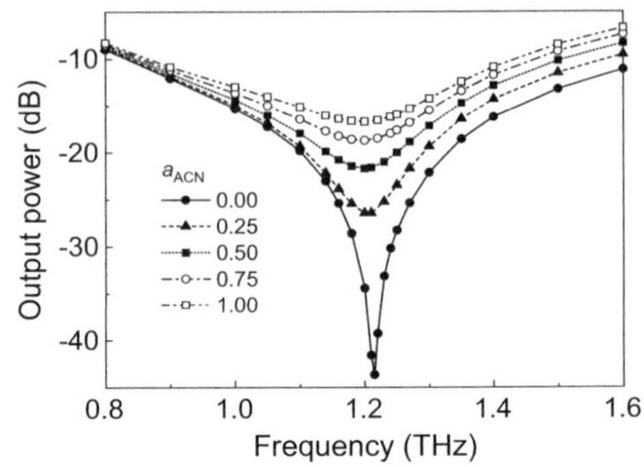

Fig. 5.13 SPR response for various volume fractions. ©[2016] IEEE. Reproduced, with permission, from [28]

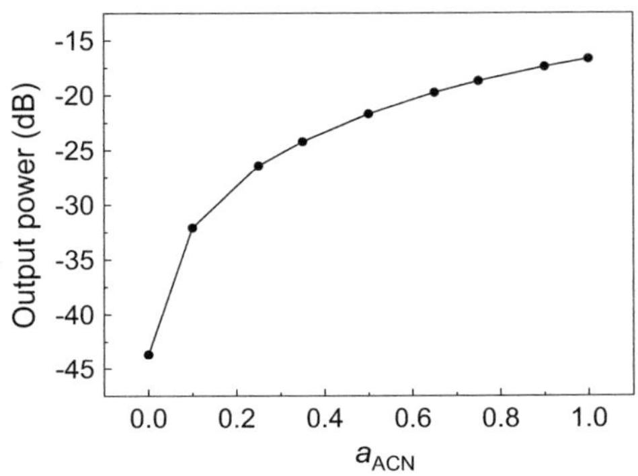

Fig. 5.14 Output power versus volume fraction. ©[2016] IEEE. Reproduced, with permission, from [28]

a concomitant increase in output power (although not shown, the amplitude of the SPP around the InSb becomes small for the reduced relative permittivity of water). Contrary to the results in Figs. 5.11 and 5.12, the change in peak frequency is quite small. This is due to the fixed temperature of $T = 293$ K.

As mentioned above, aqueous solutions decrease the absorption of THz waves. Therefore, we focus on how the output power increases with increasing volume fractions of acetonitrile. The minimum output power versus a_{ACN} is plotted in Fig. 5.14. It can be seen that the minimum power increases from -43.7 to -16.7 dB as a_{ACN} is increased from 0 to 1. It is interesting to note that the sensitivity is higher when a_{ACN} is small. Measuring the minimum output power is expected to detect the volume fractions of acetonitrile and water.

5.6 SPR Waveguide Sensor with an Analyte Container

The SPR sensors discussed above did not have a container to hold an analyte material. In this section, an SPR sensor with an analyte container is examined, with attention to the coupling of the incident field to the sensor section [42]. Figure 5.15 illustrates the SPR sensor with an analyte container, which is basically the same as in Fig. 5.2

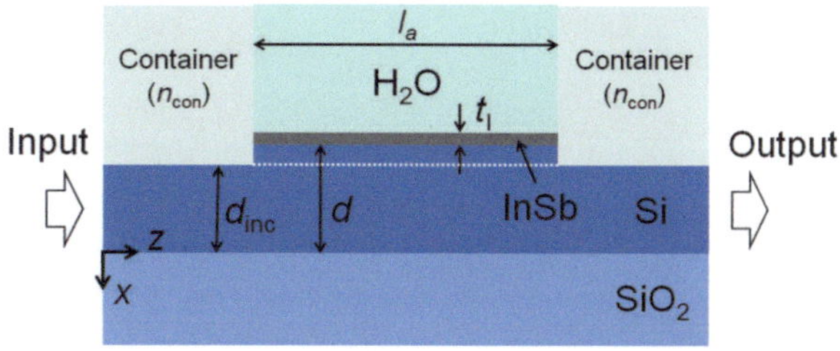

Fig. 5.15 Configuration of an SPR waveguide sensor with an analyte container. Copyright (c)2017 IEICE from [42]

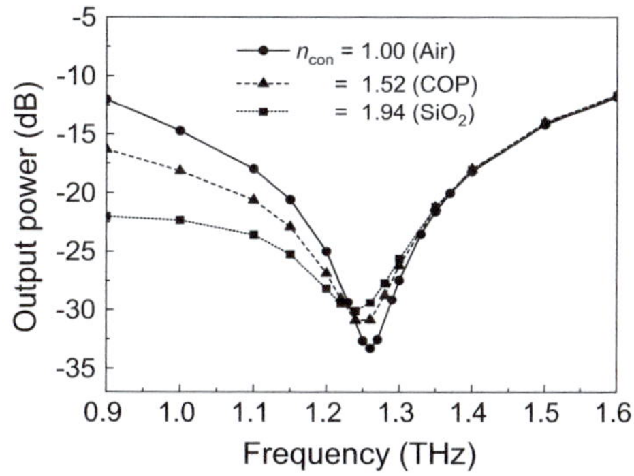

Fig. 5.16 Overlap integral. Copyright (c)2017 IEICE from [42]

except that the analyte container is placed sandwiching the sensing section and the width of the input waveguide is defined as d_{inc}.

Figure 5.16 shows the output power as a function of frequency with COP or SiO_2 selected as the analyte container, in which $d_{inc} = d$. The result without container is also included ($n_{con} = 1.00$ (air)). For practical applications, the steep frequency response is highly desirable for measurement accuracy. With the SiO_2 container, it can be seen that the output power is significantly reduced particularly on the lower frequency side, resulting in a gradual response. In contrast, the COP container suppresses the power decrease.

Next, we examine the reason why the power on the lower frequency side decreases as the refractive index of the analyte container increases. Let us now return to Fig. 5.4b, which shows the propagation loss of two surface plasmon modes excited in the sensing section. As the frequency is decreased, the propagation loss of the first mode increases while that of the second mode decreases. On the lower frequency side, since the first mode decreases rapidly with the high propagation loss, the characteristics on the lower frequency side must be determined with the excitation strength of the low-loss second mode. Then, we calculate the overlap integral between the second mode and the input eigenmode to estimate the excitation strength of the second mode. From Fig. 5.17, it can be seen that the overlap integral decreases as the refractive index of the container increases, which causes the power reduction at the lower frequency side in Fig. 5.16.

On the other hand, on the higher frequency side, the output characteristics do not depend on the refractive index of the container as shown in Fig. 5.16, which results

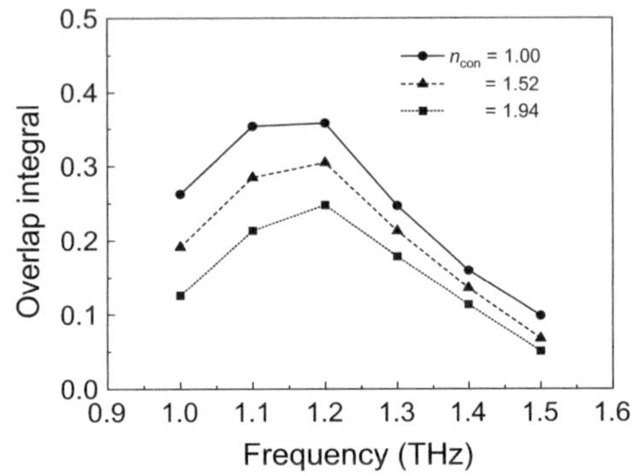

Fig. 5.17 Overlap integral for the second mode

from the excitation strength of the first mode. Figure 5.18 shows the overlap integral between the first mode and the input eigenmode. As can be seen, the overlap integral is independent of the refractive index of the container on the higher frequency side.

From the above discussion, it can be said that the container should be made of a material with a lower refractive index. As a result, we select the COP for the analyte container.

Here, we improve the frequency response of the sensor with the container by increasing the coupling strength of the second mode. The reduction of the overlap integral in Fig. 5.17 is due to the mode mismatch between the input eigenmode and the second mode. To increase the coupling, we reduce the height of the input waveguide d_{inc} as shown in Fig. 5.15. Figure 5.19 shows the overlap integral, where the reduction of d_{inc} appears to increase the overlap integral. This can be explained by the eigenmode field shown in Fig. 5.20. Compared to the field for $d_{inc} = 50\ \mu m$, the one for $d_{inc} = 40\ \mu m$ comes close to the second mode field. It is expected that the steep response of the output power is obtained even when the container is included.

Figure 5.21 shows the frequency response of the output power. It is found that the response of the sensor with the container (open circle) is almost the same as that of the sensor without the container (solid circle). Consequently, the lower-height incidence

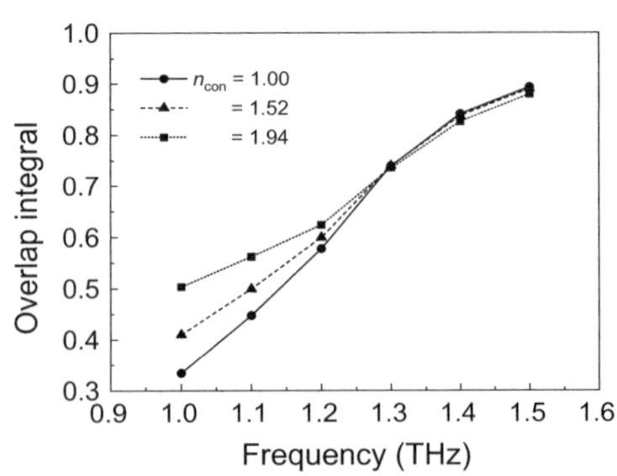

Fig. 5.18 Overlap integral for the first mode

Fig. 5.19 Overlap integral

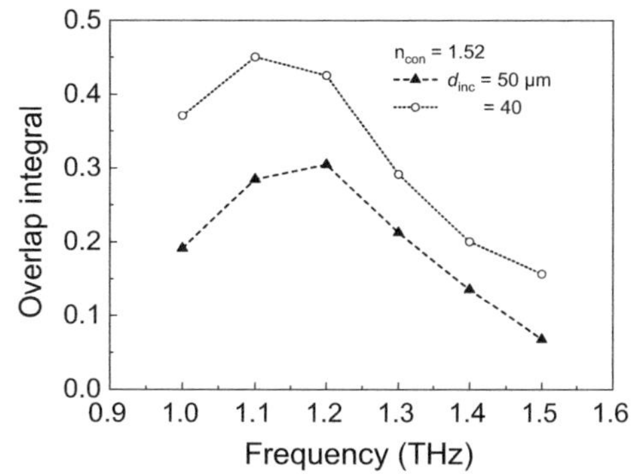

Fig. 5.20 Field distributions

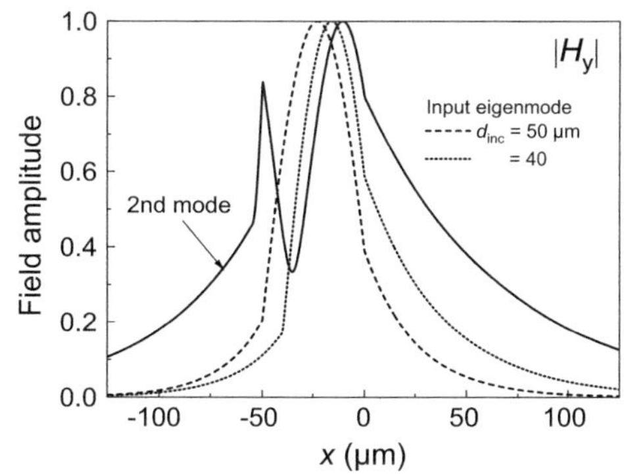

Fig. 5.21 Output power

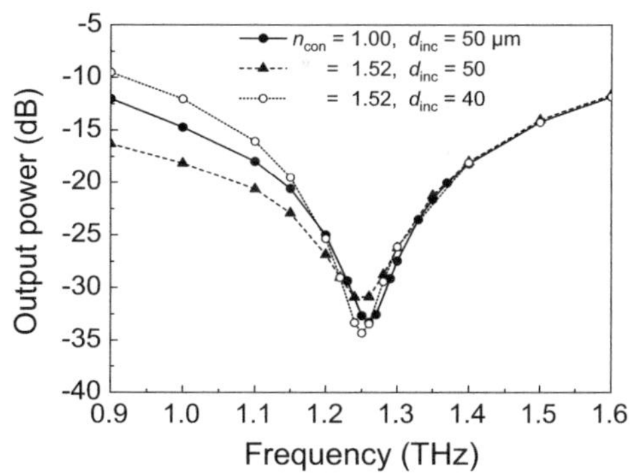

waveguide contributes to an increase in the coupling strength of the second mode, leading to a steep frequency response of the output power.

5.7 Analysis of the 3-D SPR Waveguide Sensor

In this section, we investigate a practical 3-D sensor structure shown in Fig. 5.22. The 2-D sensor structures studied so far correspond to the cross-section of the 3-D sensor shown in the right figure of Fig. 5.22. The core width is selected to be $d_W = 50\ \mu m$ and the initial structural parameters are chosen to be $w_a = 300\ \mu m$, $t_a = 150\ \mu m$ and $l_a = 400\ \mu m$. Other structural parameters are the same as those for the 2-D sensor.

Figure 5.23 shows the frequency response of the output power for $T = 300$ K. It should be noted for the results with $d_{inc} = 50\ \mu m$ that the addition of the analyte container ($n_{con} = 1.52$) causes degradation of the frequency response, as in the 2-D case shown in Fig. 5.16. On the other hand, for $d_{inc} = 40\ \mu m$, the output power even with the container is successfully increased on the lower frequency side, which improves the response better than that without the container. As a result, for the 3-D sensor, a reduction in the core height contributes to an improvement in the frequency response, as in case of the 2-D sensor. Note that the minimum output power and its frequency in Fig. 5.23 are different from those of the 2-D sensor in Fig. 5.8. Therefore, it can be said that 3-D analysis is mandatory for investigating practical 3-D sensor structures.

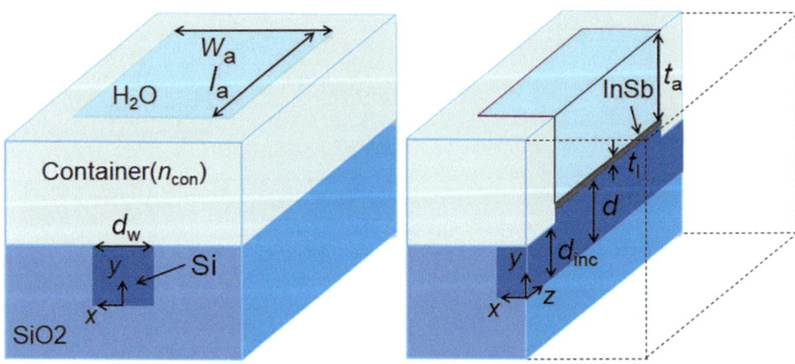

Fig. 5.22 Three-dimensional SPR waveguie sensor

Fig. 5.23 Output power

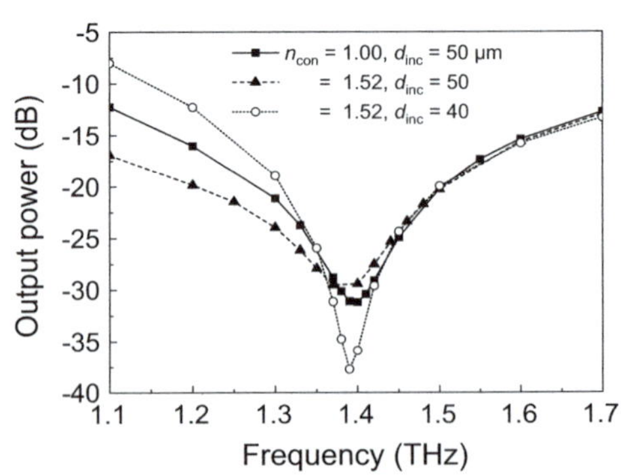

Here, we examine the effect of the size of the analyte container on the output characteristics. The output power is studied with a fixed temperature of $T = 300$ K and a fixed frequency of $f = 1.39$ THz (which is the dip frequency in Fig. 5.23). The initial structural parameters are the same as those used above: $w_a = 300$ μm, $t_a = 150$ μm and $l_a = 400$ μm. First, we study the effect of the width w_a of the container, in which the thickness is fixed to be $t_a = 150$ μm. The result is shown in Fig. 5.24. It is seen that the output power converges when the width is chosen to be more than 220 μm.

Furthermore, we study the effect of the height as shown in Fig. 5.25. It can be seen that the output power converges when the height is chosen to be more than 100 μm. So, we choose $w_a = 220$ μm and the height $t_a = 100$ μm. Next, we study the analyte length l_a corresponding to the sensing section for minimizing the output power. Figure 5.26 depicts the output power as a function of l_a. It is seen that the output power becomes minimum for $l_a = 427$ μm, where the first and second modes are out of phase. If we connect the output waveguide at $l_a = 427$ μm, we can expect a steep frequency response.

Finally, the output power for temperature is examined as the temperature changes with the structure parameters determined above: $w_a = 300$ μm, $t_a = 150$ μm and

Fig. 5.24 Output power as a function of width of the analyte container

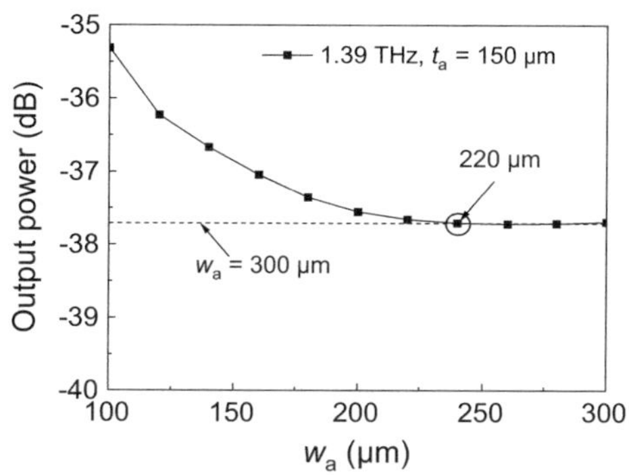

Fig. 5.25 Output power as a function of height of the analyte container

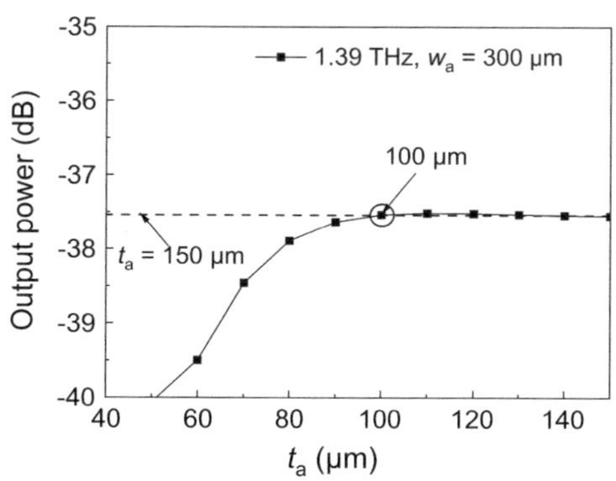

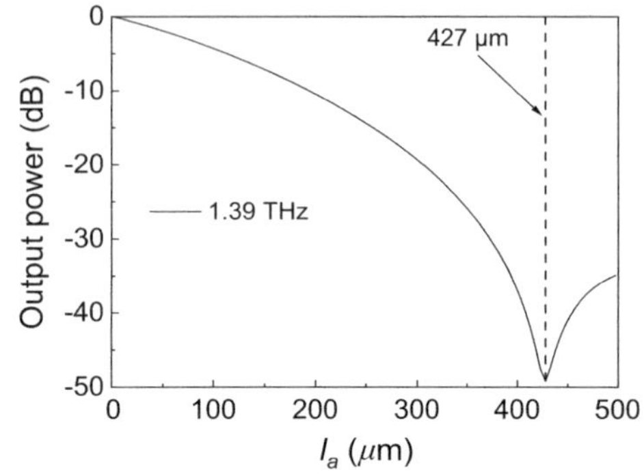

Fig. 5.26 Output power as a function of length of the analyte container

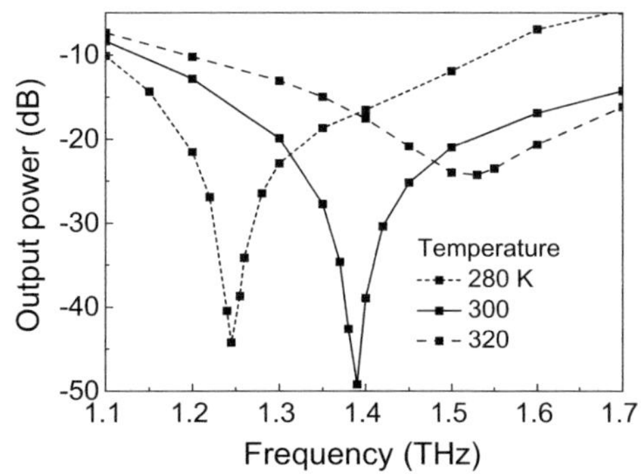

Fig. 5.27 Output power

$l_a = 427$ μm. Figure 5.27 shows the frequency response for changing the temperature from 280 to 320 K. As expected, a steep response is obtained with a design temperature of $T = 300$ K. In addition, a steep response can also be obtained at $T = 280$ K. On the other hand, the response at $T = 280$ K becomes blunt. Nevertheless, the sensitivity of the 3-D sensor is calculated to be 7.5×10^{-3} THz/K, which is almost the same as that of the 2-D sensor.

5.8 Concluding Remarks

A THz SPR waveguide sensor with a thin InSb layer adopted for the sensing section has been presented. First, the InSb and water permittivities expressed by the Drude and Debye models, respectively, are reviewed and incorporated into the frequency-dependent FDTD method. Next, the SPP eigenmodes are shown to be present in the sensing section. From the overlap integral calculations between the SPP and input eigenmodes, the SPR is expected to be maximum around 1.28 THz. The sensor is then analyzed using the FDTD method with special attention to the treatment of the InSb/water interface. The FDTD analysis shows that the SPR absorption becomes maximum at 1.26 THz, which is close to that expected from the eigenmode analysis. The SPR response is further investigated for temperature variation and shows the ability to detect the temperature of water in the range of 280 to 320 K. A mixture of an organic solvent and water is further studied, the volume fraction of which can be detected by measuring the local minimum power at the SPR frequency. Thus, an SPR waveguide sensor with a thin InSb layer can be realized even at THz frequencies.

In addition, an SPR sensor with an analyte container has been examined, with respect to the coupling of the incident field to the sensor section. A material with a lower refractive index is desirable for the analyte container. As a result, COP was chosen as the container. It has also been found that the lower-height of the incidence waveguide compared to that in the sensing section contributes to a steep frequency response of the output power.

Finally, the analysis of a practical 3-D sensor with an analyte container was performed. The size of the container is found by the full 3-D FDTD analysis. The characteristics of the 3-D sensor are found to be almost the same as the 2-D sensor. Note that since the frequency response of the 3-D sensor is different from the 2-D counterpart, we need to perform the 3-D analysis for a practical design of the 3-D sensor structures.

References

1. J. Homola, S.S. Yee, G. Gauglitz, Surface plasmon resonance sensors: review. Sens. Actuators B **54**, 3–15 (1999)
2. J. Homola, Present and future of surface plasmon resonance biosensors. Anal. Bioanalytical Chem. **377**(3), 528–539 (2003)
3. L.M. Zhang, D. Uttamchandani, Optical chemical sensing employing surface-plasmon resonance. Electron. Lett. **24**(23), 1469–1470 (1988)
4. K. Matsubara, S. Kawata, S. Minami, Optical chemical sensor based on surface-plasmon measurement. Appl. Opt. **27**(6), 1160–1163 (1988)
5. E. Kretschmann, "Die Bestimmung optischer Konstanten von Metallen durch Anregung von Oberflächenplasmaschwingungen" (in German). Z. Physik **241**(4), 313–324 (1971)
6. J. Čtyroký, J. Homola, P.V. Lambeck, S. Musa, H.J.W.M. Hoekstra, R.D. Harris, J.S. Wilkinson, B. Usievich, N.M. Lyndin, Theory and modelling of optical waveguide sensors utilising surface plasmon resonance. Sens. Actuators B **54**(1–2), 66–73 (1999)
7. J. Čtyroký, J. Homola, M. Skalsky, Tuning of spectral operation range of a waveguide surface plasmon resonance sensor. Electron. Lett. **33**(14), 1246–1248 (1997)
8. C.R. Lavers, J.S. Wilkinson, A wave-guide-coupled surface-plasmon sensor for an aqueous environment. Sens. Actuators B **22**(1), 75–81 (1994)
9. R.D. Harris, J.S. Wilkinson, Waveguide surface plasmon resonance sensors. Sens. Actuators B **29**(1–3), 261–267 (1995)
10. M.N. Weiss, R. Srivastava, H. Groger, P. Lo, S.F. Luo, A theoretical investigation of environmental monitoring using surface plasmon resonance waveguide sensors. Sens. Actuators A **51**, 211–217 (1996)
11. J. Shibayama, T. Takeuchi, N. Goto, J. Yamauchi, H. Nakano, Numerical investigation of a Kretschmann-type surface plasmon resonance waveguide sensor. J. Lightw. Technol. **25**(9), 2605–2611 (2007)
12. J. Shibayama, S. Takagi, T. Yamazaki, J. Yamauchi, H. Nakano, Numerical analysis of waveguide-based surface plasmon resonance sensor with adsorbed layer using two- and

three-dimensional beam-propagation methods. IEICE Trans. Electron. **E90-C**(1), 95–101 (2007)
13. J. Shibayama, A Kretschmann-type absorption-based surface plasmon resonance waveguide sensor. Microw. Opt. Technol. Lett. **50**(10), 2497–2500 (2008)
14. J. Shibayama, Three-dimensional numerical investigation of an improved surface plasmon resonance waveguide sensor. IEEE Photon. Technol. Lett. **22**(9), 643–645 (2010)
15. J.M. Brockman, B.P. Nelson, R.M. Corn, Surface plasmon resonance imaging measurements of ultrathin organic films. Annu. Rev. Phys. Chem. **51**, 41–63 (2000)
16. L.G. Fägerstam, Å. Frostell, R. Karlsson, M. Kullman, A. Larsson, M. Malmqvist, H. Butt, Detection of antigen-antibody interactions by surface plasmon resonance. Application to epitope mapping. J. Mol. Recognit. **3**(5–6), 208–214 (1990)
17. J. Gómez Rivas, C. Schotsch, P. Haring Bolivar, H. Kurz, Enhanced transmission of THz radiation through subwavelength holes. Phys. Rev. B **68**, 201306-1-4 (2003)
18. J. Gómez Rivas, M. Kuttge, H. Kurz, P. Haring Bolivar, J.A. Sánchez-Gil, Low-frequency active surface plasmon optics on semiconductors. Appl. Phys. Lett. **88**(8), 082106-1-3 (2006)
19. J.A. Sánchez-Gil, J. Gómez Rivas, Thermal switching of the scattering coefficients of terahertz surface plasmon polaritons impinging on a finite array of subwavelength grooves on semiconductor surfaces. Phys. Rev. B **73**(20), 205410-1-8 (2006)
20. T.H. Isaac, W.L. Barnes, E. Hendry, Determining the terahertz optical properties of subwavelength films using semiconductor surface plasmons. Appl. Phys. Lett. **93**(24), 241115-1-3 (2008)
21. M.K. Chen, Y.C. Chang, C.E. Yang, Y. Guo, J. Mazurowski, S. Yin, P. Ruffin, C. Brantley, E. Edwards, C. Luo, Tunable terahertz plasmonic lenses based on semiconductor microslits. Microw. Opt. Tech. Lett. **52**(4), 979–981 (2010)
22. B. Hu, Q.J. Wang, S.W. Kok, Y. Zhang, Active focal length control of terahertz slitted plane lenses by magnetoplasmons. Plasmonics **7**(2), 191–199 (2012)
23. B. Hu, Q.J. Wang, Y. Zhang, Broadly tunable one-way terahertz plasmonic waveguide based on nonreciprocal surface magneto plasmons. Opt. Lett. **37**(11), 1895–1897 (2012)
24. J. Tao, B. Hu, X.Y. He, Q.J. Wang, Tunable subwavelength terahertz plasmonic stub waveguide filters. IEEE Trans. Nanotech. **12**(6), 1191–1197 (2013)
25. J. Shibayama, Y. Wada, J. Yamauchi, H. Nakano, Application of the explicit and implicit FDTD methods to the analysis of a terahertz plasmonic grating, in *Progress in Electromagnetics Research Symposium* (Guangzhou, China, 2014), p. 1922
26. Q. Wang, Q. Tang, D. Zhang, Z. Wang, Y. Huang, Tunable terahertz spectral filter based on temperature controlled subwavelength InSb grating. Superlattice. Microst. **75**, 955–961 (2014)
27. H. Liu, G. Ren, Y. Gao, Y. Lian, Y. Qi, S. Jian, Tunable subwavelength terahertz plasmon-induced transparency in the InSb slot waveguide side-coupled with two stub resonators. Appl. Opt. **54**(13), 3918–3924 (2015)
28. J. Shibayama, K. Shimizu, J. Yamauchi, H. Nakano, Surface plasmon resonance waveguide sensor in the terahertz regime. IEEE/OSA J. Lightw. Technol. **34**(10), 2518–2525 (2016)
29. H.J. Liebe, G.A. Hufford, T. Manabe, A model for the complex permittivity of water at frequencies below 1 THz. Int. J. Infrared Millimeter Waves **12**(7), 659–675 (1991)
30. D. Yevick, B. Hermansson, New formulations of the matrix beam propagation method: Application to rib waveguides. IEEE J. Quantum Electron. **25**(2), 221–229 (1989)
31. J. Shibayama, T. Yamazaki, J. Yamauchi, H. Nakano, Eigenmode analysis of a light-guiding metal line loaded on a dielectric substrate using the imaginary-distance beam-propagation method. J. Lightw. Technol. **23**(3), 1533–1539 (2005)
32. D. Grischkowsky, S. Keiding, M. van Exter, Ch. Fattinger, Far-infrared time-domain spectroscopy with terahertz beams of dielectrics and semiconductors. J. Opt. Soc. Am. B **7**(10), 2006–2015 (1990)
33. Y.H. Avetisyan, H.S. Hakobyan, V.R. Tadevosyan, Analysis of plasmon terahertz waveguide formed by a dielectric ridge on semiconductor surface. J. Contemp. Phy. **46**(5), 361–367 (2011)
34. T. Arima, T. Uno, Boundary treatment method of dispersive materials analysis by FDTD method (in Japanese). Trans. IEICE **J91-B**(9), 1066–1068 (2008)

35. Y. Zhao, P. Belov, Y. Hao, Accurate modeling of the optical properties of left-handed media using a finite-difference time-domain method. Phys. Rev. E **75**(3), 037602-1-4 (2007)
36. J. Shibayama, R. Ando, A. Nomura, J. Yamauchi, H. Nakano, Simple trapezoidal recursive convolution technique for the frequency-dependent FDTD analysis of a Drude-Lorentz model. IEEE Photon. Technol. Lett. **21**(2), 100–102 (2009)
37. T. Yoshimura, J. Yamauchi, J. Shibayama, H. Nakano, Three-dimensional BPM analysis of a Kretschmann-type SPR sensor with an analyte container (in Japanese), in *Proceedings of the IEICE Social Conference*, C-3-24, p. 141 (2012)
38. J.J. Ramsden, S.E. Webber, M. Graetzel, Luminescence of colloidal cadmium sulfide particles in acetonitrile and acetonitrile/water mixtures. J. Phys. Chem. **89**(13), 2740–2743 (1985)
39. D.S. Venables, C.A. Schmuttenmaer, Far-infrared spectra and associated dynamics in acetonitrile-water mixtures measured with femtosecond THz pulse spectroscopy. J. Chem. Phys. **108**(12), 4935–4944 (1998)
40. T. Takamuku, M. Tabata, A. Yamaguchi, J. Nishimoto, M. Kumamoto, H. Wakita, T. Yamaguchi, Liquid structure of acetonitrile-water mixtures by x-ray diffraction and infrared spectroscopy. J. Phys. Chem. B **102**(44), 8880–8888 (1998)
41. N.Y. Tan, R. Li, P. Bräuer, C. D'Agostino, L.F. Gladden, J.A. Zeitler, Probing hydrogen-bonding in binary liquid mixtures with terahertz time-domain spectroscopy: a comparison of Debye and absorption analysis. Phys. Chem. **17**, 5999–6008 (2015)
42. J. Shibayama, K. Shimizu, J. Yamauchi, H. Nakano, FDTD analysis of the terahertz SPR waveguide sensor with an analyte container. IEICE Trans. Electron **J100-C**(5), 239–241 (2017). (Japanese Edition)

Chapter 6
Kretschmann- and Otto-Type SPR Waveguide Sensors at Terahertz Frequencies

Abstract This chapter presents numerical studies of Kretschmann- and Otto-type surface plasmon resonance (SPR) waveguide sensors at terahertz frequencies. The reflectance of a three-layer model for the Kretschmann sensor is studied using the Fresnel equation to determine the angle of incidence of the waveguide. The frequency-dependent finite-difference time-domain method is used to study the SPR response of the waveguide sensor to temperature variations. It is found that a high sensitivity of 2.63×10^{-2} THz/K, more than three times that of the conventional SPR waveguide sensor, is successfully achieved with a sensing length about half that of the conventional one. The Otto-type waveguide sensor is also investigated and discussed. It is shown that the sensor provides steep and stable SPR responses while maintaining the sensitivity of the Kretschmann type.

6.1 Introduction

It is well known that surface plasmon resonance (SPR) sensors at optical frequencies have been commercially available [1]. The SPR sensors have been mainly used to detect biochemical reactions, e.g. protein-DNA interactions, antigen-antibody complex reactions, etc. Basically, the SPR sensors are based on the so-called Kretschmann configuration [2], in which a prism coated with a thin metal layer is adopted. Unfortunately, this configuration is not suitable for monolithic integration into optical circuits. To overcome this problem, several SPR waveguide-based sensors have been investigated [3–7]. SPR sensors based on metal-insulator-metal and insulator-metal-insulator waveguides have also been investigated [8–10].

At terahertz (THz) frequencies, vibrational modes of hydrogen exist around 1 THz [11]. This fact has motivated researchers to investigate SPR sensors at THz frequencies. We have also investigated an SPR waveguide sensor in which a thin indium antimonide (InSb) layer has been adopted in the sensing section [12, 13]. Several SPR waveguide sensors with InSb components have also been investigated recently [14–16]. Note that the waveguide sensors are generally less sensitive to the variation of the refractive index compared to the Kretschman-based SPR sensors.

J. Shibayama, *FDTD Analysis of Plasmonic Devices*, Springer Series in Optical Sciences 254, https://doi.org/10.1007/978-981-96-8130-3_6

In this chapter, SPR waveguide sensors with higher sensitivity at THz frequencies are numerically investigated [17]. In particular, we introduce the Kretschmann [2] and Otto [18] configurations into THz waveguide sensors. Using the three-layer mode of the Kretschmann waveguide sensor, we first choose the angle of incidence of the waveguide to be 45°. Then, using the frequency-dependent finite-difference time-domain (FDTD) method based on the trapezoidal recursive convolution (TRC) technique, the frequency response of the Kretschmann type sensor is investigated for temperature variations. It is found that the sensor has a high sensitivity of 2.63×10^{-2} THz/K, which is more than three times that of the conventional SPR waveguide sensor [12, 15]. In addition, the sensing length of the present sensor is reduced to about half of that of the conventional sensor. Note that the shape of the SPR response depends on the temperature due to the effect of skin depth. Finally, we investigate an Otto-type waveguide sensor in which the skin depth effect on the SPR response can be reduced. It is shown that the Otto-type sensor provides steep and stable SPR responses for temperature variations without losing sensitivity compared to the Kretschmann type.

6.2 Kretschmann-Type Waveguide Sensor

The configuration of the Kretschmann waveguide sensor is shown in Fig. 6.1. The core and substrate are made of SiO_2 and polyethylene (PE), respectively, whose relative permittivities are set to 1.942^2 and 1.52^2 (refractive index squared). The core width of the waveguide is chosen to be W=100 μ, and the angle of incidence of the waveguide is defined as θ. A thin InSb layer of $t = 2\ \mu\text{m}$ is assumed for the sensing section and water is given as the analyte outside the InSb layer. The permittivities of InSb and water are expressed using the Drude and double Debye models, which are given in Cap. 6.2.

To estimate the reflectivity of the sensor, we reduce the waveguide section 1 in Fig. 6.1 to a homogeneous medium, resulting in a three-layer model. The relative

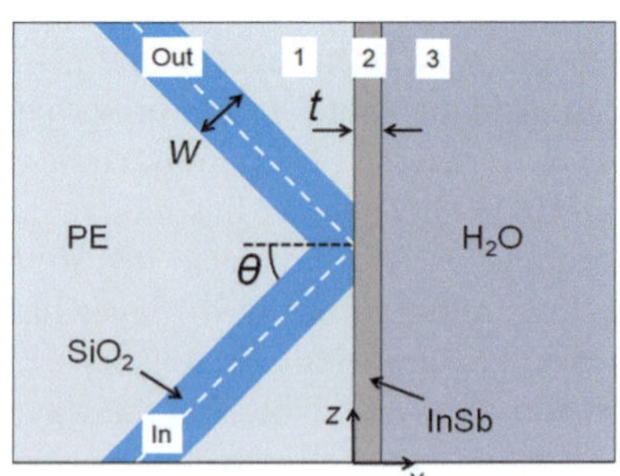

Fig. 6.1 Configuration of Kretschmann-type waveguide sensor. From [17] Copyright ©2020 by John Wiley Sons, Inc. Reproduced by permission of John Wiley & Sons, Inc

permittivity of the waveguide section is approximated to be the effective permittivity of the fundamental TM mode as a function of frequency. The reflectivity rsp of the three-layer model is calculated by the following Fresnel equation [19]:

$$r_{SP} = \left| \frac{r_{12} + r_{23} \exp(i2k_{2z}t)}{1 + r_{12}r_{23} \exp(i2k_{2z}t)} \right|^2 \quad (6.1)$$

where

$$r_{ij}\& = \frac{\varepsilon_i k_{jz} - \varepsilon_j k_{iz}}{\varepsilon_i k_{jz} + \varepsilon_j k_{iz}}$$

$$k_{iz}\& = \frac{\omega}{c} \left(\varepsilon_i - \varepsilon_1 \sin^2 \theta_1\right)^{1/2}$$

in which ε_i is the dielectric constant of the ith medium ($i = 1, 2, 3$), θ is the incidence angle of the plane wave and c is the speed of light in a vacuum. Figure 6.2 shows the reflectivity as a function of incidence angle and frequency. It is seen that the reflectivity becomes minimum at around 45°, resulting in a dip frequency of 2.5 THz. As a result, we set the incidence angle of the waveguide is set to be $\theta = 45°$.

To handle the dispersion materials of the waveguide sensor, we use the frequency-dependent FDTD method based on the TRC technique, which is simpler than the commonly used piecewise linear recursive convolution technique. The averaged dispersion model is derived separately along the interface between the InSb layer and the water for stable calculations [12]. Along the other boundaries, we impose an averaged permittivity between the two materials to satisfy the boundary condition

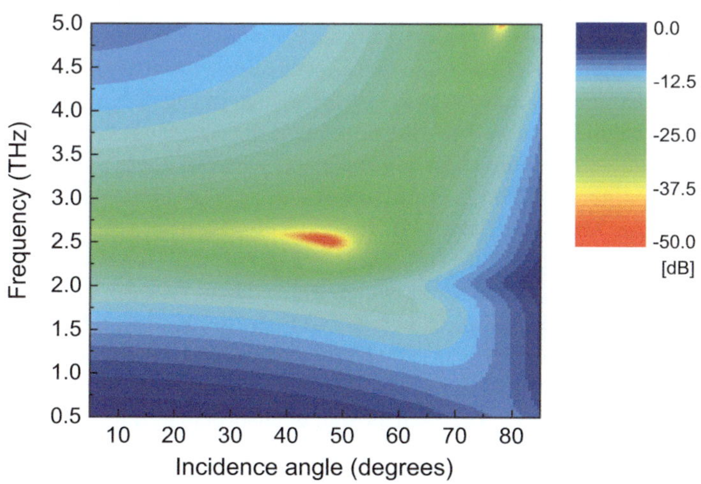

Fig. 6.2 Reflectivity as a function of incidence angle and frequency. From [17] Copyright ©2020 by John Wiley Sons, Inc. Reproduced by permission of John Wiley & Sons, Inc

Fig. 6.3 Output power. From [17] Copyright ©2020 by John Wiley Sons, Inc. Reproduced by permission of John Wiley & Sons, Inc

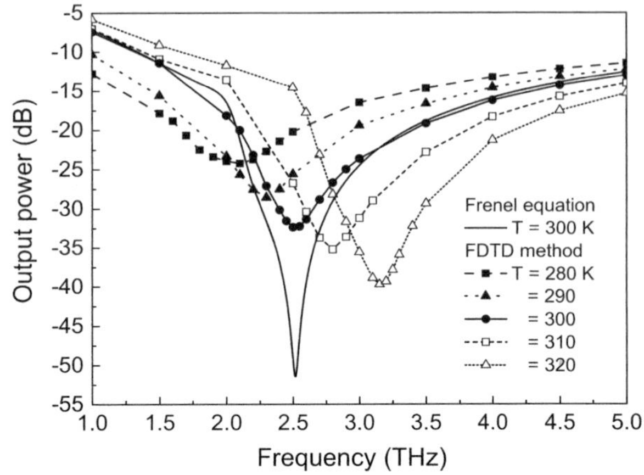

of continuity of the tangential electric field. The fundamental TM mode is launched from the input waveguide and the power is observed at the output waveguide. Outside the computational domain, we set the convolutional perfectly matched layer to absorb the outgoing waves.

The output power versus frequency is shown in Fig. 6.3 for various temperatures. For reference, the reflectivity of the three-layer model calculated with (6.1) is included at T = 300 K. It can be seen that the dip frequency of the waveguide sensor is almost the same as that of the three-layer model at T = 300 K. However, the absolute magnitude of the response signal at T = 300 K is larger and the response is broader than that of the three-layer model. This is because the wave incident on the InSb layer has a finite spot size of the guided mode of the waveguide compared to the plane wave case. Although not shown, as the core width is increased, the frequency response becomes close to that of the three-layer case.

We further discuss the temperature characteristics in Fig. 6.3, where the temperature is varied from 280 to 320 K. It is noted that the frequency dip shifts to higher frequencies as the temperature is increased. Here, we investigate the sensitivity with respect to the temperature change (not with respect to the refractive index at optical frequencies [20]). The sensitivity is defined as $\Delta f/\Delta K$, where Δf is the frequency shift and ΔK is the temperature shift [12]. In Fig. 6.3, a high sensitivity of 2.63×10^{-2} THz/K is successfully achieved, which is more than three times that of the conventional waveguide sensor, e.g., 7.5×10^{-3} THz/K in [12] and 6.6×10^{-3} THz/K in [15]. This is because the SPR is effectively excited in the Kretschmann type sensor. Figure 6.4 shows the field distributions at the off and on resonance frequencies at T = 300 K. At 2.5 THz, the input power is absorbed in the analyte due to the SPR excitation. It is observed that the localization of the THz wave in the analyte is limited to the region near the waveguide core. The interaction between the THz wave and the analyte takes place within a length of about 200 μm, which is almost half the length of the conventional waveguide sensors [12, 15].

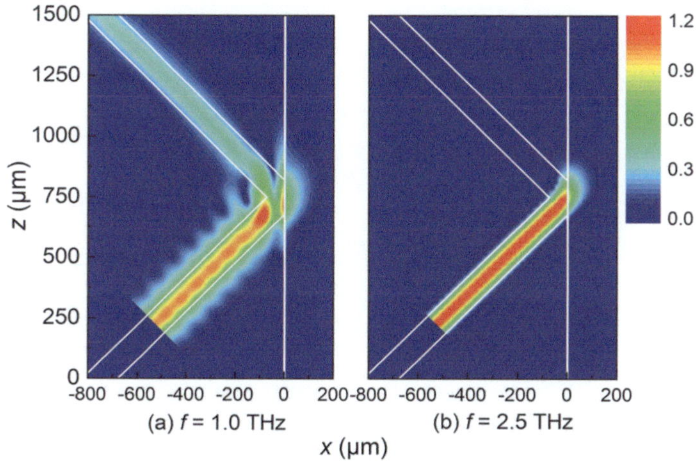

Fig. 6.4 Field distributions. **a** 1 THz and **b** 2.5 THz. From [17] Copyright ©2020 by John Wiley Sons, Inc. Reproduced by permission of John Wiley & Sons, Inc

Note that the shape of the SPR response in Fig. 6.3 is temperature dependent. This temperature dependence is due to a skin depth of the THz waves and a finite, constant thickness of the InSb layer assumed in the Kretschmann type sensor. Here we estimate the skin depth δ of InSb using the well-known formula: $\delta = \sqrt{2/(\omega\mu\sigma)}$where μ is the permeability and σ is the conductivity calculated from the imaginary part of the permittivity of InSb. For example, at 2.5 THz, the skin depth is estimated to be 35 to 19 μm when the temperature is varied from 280 to 320 K.

6.3 Otto-Type Waveguide Sensor

As discussed above, the SPR response of the Kretschmann-type sensor shows temperature dependence, due to the constant thickness of the InSb layer. In contrast, an arbitrarily thick InSb layer can be used for the Otto configuration. Therefore, we expect to reduce the temperature dependence of the shape of the SPR response. The configuration of the Otto-type waveguide sensor with a thick InSb layer is shown in Fig. 6.5, where the InSb and water layers in Fig. 6.1 are replaced with each other.

Figure 6.6 shows the reflectivity as a function of incidence angle and frequency using (6.1). It can be seen that frequency dip can be found at around 60° and 5 THz. However, to obtain an SPR dip frequency close to that of the Kretschmann type sensor, we choose the same angle of incidence of the waveguide, i.e., $\theta = 45°$.

The output power from the waveguide is shown in Fig. 6.7 for various temperatures. As expected, it is found that the response maintains its shape regardless of the temperature when compared with the Kretschmann type sensor in Fig. 6.3. This is due to the fact that the thick InSb layer removes the effect of the skin depth for

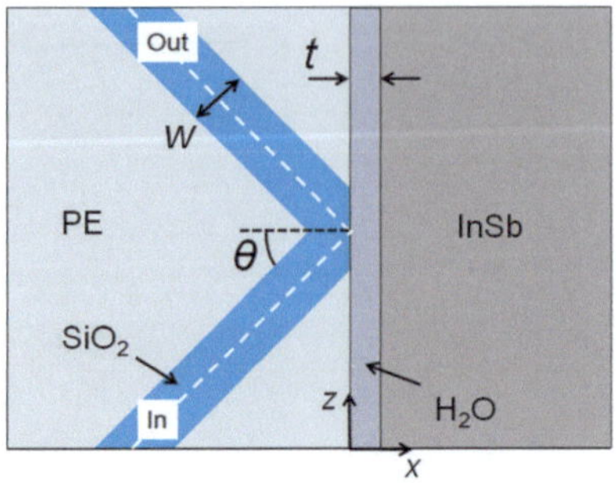

Fig. 6.5 Configuration of Otto-type waveguide sensor. From [17] Copyright ©2020 by John Wiley Sons, Inc. Reproduced by permission of John Wiley & Sons, Inc

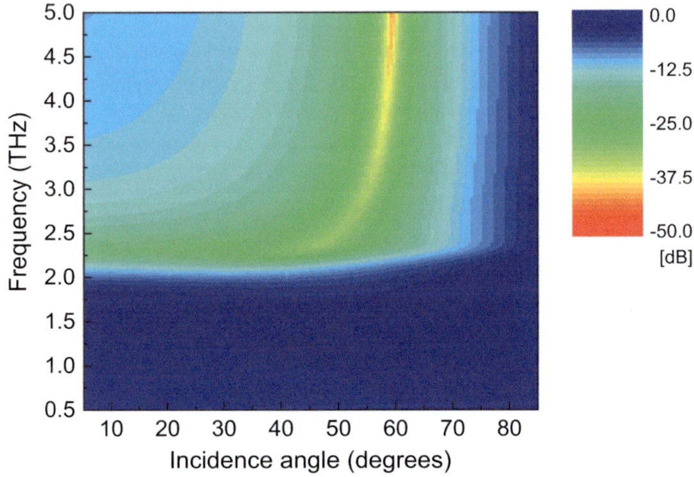

Fig. 6.6 Reflectivity as a function of incidence angle and frequency

a thin InSb one. It is also worth mentioning that the sensitivity is calculated to be 2.83×10^{-2}, which is 8% higher than that of the Kretschmann type sensor, without increasing the sensing length by about 200 μm.

It should be also noted in Fig. 6.7 that the output power at $T = 300$ K for the waveguide sensor is smaller than that calculated with the Fresnel equation. Figure 6.8 illustrates the field distributions at the off and on resonance frequencies at T = 300 K. At 2.3 THz, the radiation field is not coupled to the output waveguide. This is probably the reason why the output power of the waveguide sensor is small.

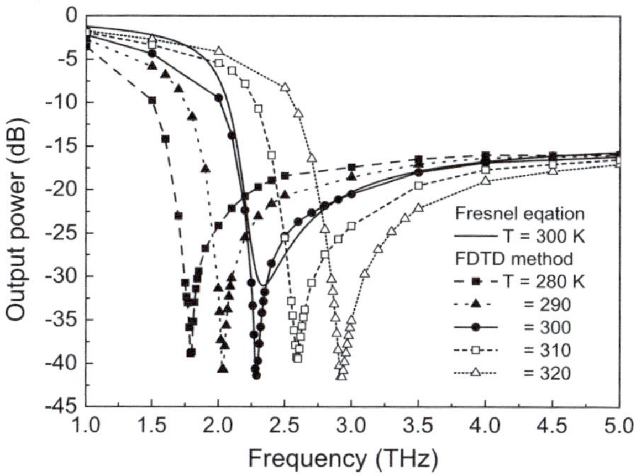

Fig. 6.7 Output power. From [17] Copyright ©2020 by John Wiley Sons, Inc. Reproduced by permission of John Wiley & Sons, Inc

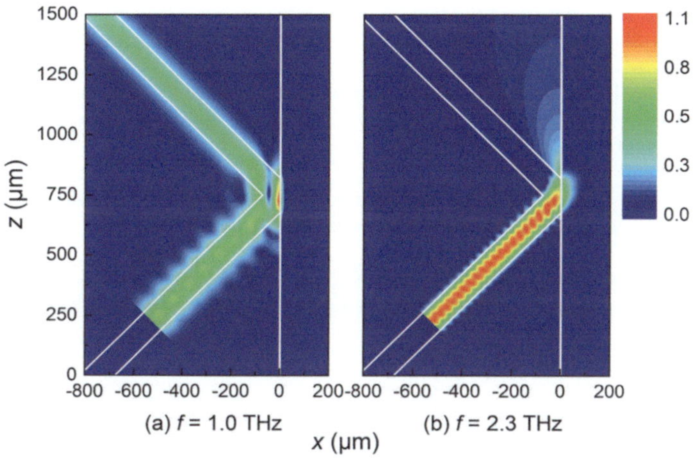

Fig. 6.8 Field distributions. **a** 1 THz and **b** 2.3 THz

6.4 Concluding Remarks

We have numerically investigated Kretschmann and Otto type SPR waveguide sensors at terahertz frequencies. A three-layer model of the Kretschmann sensor is used to determine the angle of incidence of the waveguide. The sensing performance of the sensor is investigated using the frequency-dependent FDTD method based on the simple TRC technique. It is found that a high sensitivity (2.83×10^{-2}) more than three times that of the conventional SPR waveguide sensor is successfully achieved with a sensing length of about 200 μm, which is half of the conventional one. The Otto-type SPR waveguide sensor, in which steep and stable SPR responses are obtained, is also investigated and discussed. The three-dimensional analysis of the waveguide sensors is left for future study.

References

1. J. Homola, S.S. Yee, G. Gauglitz, Surface plasmon resonance sensors: review. Sens. Actuators B **54**(1), 3–15 (1999)
2. E. Kretschmann, H. Raether, Radiative decay of non radiative surface plasmons excited by light. Z. Naturforsch **23**(12), 2135–2136 (1968)
3. J. Čtyroký, J. Homola, P.V. Lambeck, S. Musa, H.J.W.M. Hoekstra, R.D. Harris, J.S. Wilkinson, B. Usievich, N.M. Lyndin, Theory and modelling of optical waveguide sensors utilising surface plasmon resonance. Sens. Actuators B **54**(1–2), 66–73 (1999)
4. C.R. Lavers, J.S. Wilkinson, A wave-guide-coupled surface-plasmon sensor for an aqueous environment. Sens. Actuators B **22**(1), 75–81 (1994)
5. R.D. Harris, J.S. Wilkinson, Waveguide surface plasmon resonance sensors. Sens. Actuators B **29**(1–3), 261–267 (1995)
6. J. Shibayama, T. Takeuchi, N. Goto, J. Yamauchi, H. Nakano, Numerical investigation of a Kretschmann-type surface plasmon resonance waveguide sensor. J. Lightw. Technol. **25**(9), 2605–2611 (2007)
7. J. Shibayama, S. Takagi, T. Yamazaki, J. Yamauchi, H. Nakano, Numerical analysis of waveguide-based surface plasmon resonance sensor with adsorbed layer using two- and three-dimensional beam-propagation methods. IEICE Trans. Electron. **E90-C**(1), 95–101 (2007)
8. S. Zhan, Y. Peng, Z. He, B. Li, Z. Chen, H. Xu, H. Li, Tunable nanoplasmonic sensor based on the asymmetric degree of Fano resonance in MDM waveguide. Sci Rep. **6**(22428), 1–8 (2016)
9. B.X. Li, H.J. Li, L.L. Zeng, S.P. Zhan, Z.H. He, Sensing application in Fano resonance with T-shape structure. J Lightw Technol. **34**(14), 3342–3347 (2016)
10. J. Zhu, N. Li, Novel high sensitivity SPR sensor based on surface plasmon resonance technology and IMI waveguide structure. Results Phys. **17**(103049), 1–9 (2020)
11. M. Tonouchi, Cutting-edge terahertz technology. Nature Photon. **1**, 97–105 (2007)
12. J. Shibayama, K. Shimizu, J. Yamauchi, H. Nakano, Surface plasmon resonance waveguide sensor in the terahertz regime. J. Lightw. Technol. **34**(10), 2518–2525 (2016)
13. J. Shibayama, A Kretschmann-type absorption-based surface plasmon resonance waveguide sensor. Microw. Opt. Technol. Lett. **50**(10), 2497–2500 (2008)
14. Y. Ma, J. Zhou, J. Pištora, M. Eldlio, N. Nguyen-Huu, H. Maeda, Q. Wu, M. Cada, Subwavelength InSb-based slot wavguides for THz transport: concept and practical implementations. Sci. Rep. **6**, 38784-1-10 (2016)
15. Y. Ma, N. Nguyen-Huu, J. Zhou, H. Maeda, Q. Wu, M. Eldlio, J. Pštora, M. Cada, Mach-Zehnder interferometer-based integrated terahertz temperature sensor. IEEE J. Sel. Top. Q. Electron. **23**(4), 4601607-1-8 (2017)
16. J. Chochol, K. Postava, M. Cada, J. Pištora, Experimental demonstration of magnetoplasmon polariton at InSb(InAs)/dielectric interface for terahertz sensor application. Sci. Rep. **7**, 13117-1-8 (2017)
17. J. Shibayama, K. Mitsutake, J. Yamauchi, H. Nakano, Kretschmann- and Otto-type surface plasmon resonance waveguide sensors in the terahertz regime. Microw. Opt. Tech. Lett. **63**(1), 103–106 (2021)

18. A. Otto, Excitation of nonradiative surface plasma waves in silver by the method of frustrated total reflection. Z. Phys. **216**(4), 398–410 (1968)
19. R.V. Andaloro, H.J. Simon, R.T. Deck, Temporal pulse reshaping with surface waves. Appl. Opt. **33**(27), 6340–6347 (1994)
20. C. Rizal, V. Belotelov, D. Ignatyeva, A.K. Zvezdin, S. Pisana, Surface plasmon resonance (SPR) to magneto-optic SPR. Condens. Matter **4**(50), 1–7 (2019)

Chapter 7
Waveguide Polarizers at Terahertz Frequencies

Abstract This chapter presents an efficient analysis of TM-pass/TE-stop THz waveguide polarizers using a hybrid-implicit-explicit finite-difference time-domain (HIE-FDTD) method. A two-dimensional HIE-FDTD method using the trapezoidal recursive convolution (TRC) technique is newly developed to analyze waveguide polarizers. The maximum error of the guided mode power is 0.81%, and the computation time can be reduced to about 36% of that of the traditional explicit FDTD method when the ratio of the spatial sampling widths in the two directions is 1:16. The TE mode propagation is also calculated and the extinction ratio characteristics of 2-D polarizers are clarified. In addition, a thee-dimensional HIE-FDTD method is developed to analyze practical 3-D polarizers. The computation time can be reduced to about 22% of that of the traditional explicit FDTD method when the ratio of the spatial sampling widths in the three directions is 1:20:20. It is shown that an extinction ratio of more than 25 dB is obtained with a polarizer length of 126 μm. The necessity of the TRC technique for analysis using the HIE-FDTD method with a large CFLN is also discussed. By comparing the 2-D and 3-D results, it is also shown that 3-D analysis is necessary for the study of practical polarizers. The final study is another polarizer with InAs sidewalls placed outside the dielectric buffer layers on either side of the core. The TE waves generate surface plasmons and become lossy, so the field is attenuated and only TM modes are guided. Also consider the case where the dielectric buffer layer is chosen to be hollow. When the buffer is hollow, it becomes a waveguide for TE and TM mode transmission, but when water is injected into the buffer, the waveguide is found to act as a TE stop/TM pass polarizer.

7.1 Introduction

Several waveguide devices in the terahertz (THz) band have been investigated so far. For example, a surface plasmon resonance (SPR) sensor using indium antimonide (InSb) [1], which is capable of guiding surface plasmon polaritons (SPP) in the THz band, was discussed in Chaps. 6 and 7. Since the TM wave is required to excite the SPP in the sensor, polarizers for TM transmission and TE rejection are required at the front end of the sensor. Therefore, we have also proposed waveguide-type THz

J. Shibayama, *FDTD Analysis of Plasmonic Devices*, Springer Series in Optical Sciences 254, https://doi.org/10.1007/978-981-96-8130-3_7

polarizers using indium arsenide (InAs), which has low temperature dependence. It should be noted that both the SPR sensor and the polarizer utilize thin InSb and InAs films, which require fine spatial meshes in the thickness direction of the films. As a result, the explicit FDTD method requires a small time step under the CFL condition, which may increase the computation time. To alleviate this problem, the use of the HIE-FDTD method [2–6], which uses the implicit method only in the fine direction, is quite favorable.

In this chapter, the two-dimensional (2-D) HIE-FDTD method is used to efficiently analyze a TM-pass/TE-stop THz waveguide polarizers with a thin InAs layer [7]. First, we extend the HIE-FDTD method to a frequency-dependent version by introducing the trapezoidal recursive convolution (TRC) technique. When the ratio of the spatial increments in the two directions is 1:16, a maximum error in the guided-mode power of 0.81% is obtained, and the computational time can be reduced to about 36% of that of the explicit FDTD method. The propagation of TE modes is also analyzed and the extinction ratio characteristics of polarizers are discussed.

Furthermore, a three-dimensional (3-D) polarizer is treated as a practical application [8]. A 3-D frequency-dependent HIE-FDTD method is also developed using the TRC technique. The field distributions, guided-mode power and extinction ratio are calculated and compared with those obtained by the traditional explicit FDTD method, showing excellent agreement between the results of the two methods. The computation time can be reduced to about 22% of the traditional explicit FDTD method. It is also shown that an extinction ratio of more than 25 dB is obtained with a polarizer length of 126 μm. Through comparing the 2-D and 3-D results, it is also shown that 3-D analysis is necessary for the study of practical polarizers.

The final study is another polarizer with InAs sidewalls placed outside the dielectric buffer layers on both sides of the core, which is analyzed by the developed 3-D HIE-FDTD method. For the presented polarizer, the TE waves generate surface plasmons and become lossy, so the field is attenuated and only TM modes are guided. It follows that the polarizer is not sensitive to the device length. Also consider the case where the dielectric buffer layers are chosen to be hollow. When the buffers are hollow, the waveguide results in TE and TM mode transmission. In contrast, when water is injected into the buffer, the waveguide acts as a TE stop/TM pass polarizer.

7.2 Formulation of the 2-D Frequency-Dependent HIE-FDTD Method

At THz frequencies InAs is considered as a frequency-dependent material and its dispersion property is expressed by the following Drude model:

$$\varepsilon_r(\omega) = \varepsilon_\infty + \frac{\omega_{\mathrm{D}}^2}{j\omega\nu_{\mathrm{D}} - \omega^2} \tag{7.1}$$

This model is the same as (3.1) except that the Lorentz term is omitted. With the TRC technique, the frequency-dependent HIE-FDTD method is derived as

$$E_x^{n+1/2} = \frac{\varepsilon_\infty - \frac{\chi^0}{2}}{\varepsilon_\infty + \frac{\chi^0}{2}} E_x^{n-1/2} + \frac{1}{\varepsilon_\infty + \frac{\chi^0}{2}} \phi_x^{n-1/2} - \frac{c\Delta t}{\varepsilon_\infty + \frac{\chi^0}{2}} \frac{\partial H_y^n}{\partial z} \tag{7.2}$$

$$E_z^{n+1} = \frac{\varepsilon_\infty - \frac{\chi^0}{2}}{\varepsilon_\infty + \frac{\chi^0}{2}} E_z^{n} + \frac{1}{\varepsilon_\infty + \frac{\chi^0}{2}} \phi_z^{n} + \frac{c\Delta t}{\varepsilon_\infty + \frac{\chi^0}{2}} \frac{\partial}{\partial x} \left(\frac{H_y^{n+1} + H_y^n}{2} \right) \tag{7.3}$$

$$H_y^{n+1} = H_y^n - \frac{c\Delta t}{\mu_r} \left[\frac{\partial E_x^{n+1/2}}{\partial z} - \frac{\partial}{\partial x} \left(\frac{E_z^{n+1} + E_z^n}{2} \right) \right] \tag{7.4}$$

where

$$\chi^0 = \frac{\omega_{\mathrm{D}}^2}{\nu_{\mathrm{D}}} \left[\Delta t - \frac{1}{\nu_{\mathrm{D}}} (1 - e^{-\nu_{\mathrm{D}} \Delta t}) \right]$$

$$\phi_\xi^n = \frac{E_\xi^n + E_\xi^{n-1}}{2} \Delta\chi^0 + e^{-\nu_{\mathrm{D}} \Delta t} \phi_\xi^{n-1}$$

$$\Delta\chi^0 = -\frac{\omega_{\mathrm{D}}^2}{\nu_{\mathrm{D}}^2} (1 - e^{-\nu_{\mathrm{D}} \Delta t})^2$$

in which $\xi = x$ or z.

In the above formulation, we apply the implicit method only in the x-direction because we use small spatial sampling widths in the x-direction (corresponding to the thickness direction of the polarizer). That is, for the derivative in the x direction, we take the average of the derivatives as in (7.3) and (7.4). First, (7.2) is solved explicitly. Then, substitute (7.4) into (7.3) and solve implicitly, followed by solving (7.4) explicitly.

The CFL conditions for the HIE-FDTD method is expressed as

$$\Delta t_{\mathrm{Semi}} \leq \frac{n_{\min}}{c\sqrt{(1/\Delta z)^2}} \tag{7.5}$$

where $n_{\min}$ is the minimum value of the refractive index in the calculation region. Since the implicit calculation removes the condition in the x direction, the condition is only determined by the mesh size in the z direction. The ratio of the time step size between the 2-D HIE-FDTD and explicit FDTD methods is defined as CFLN $= \Delta t_{\mathrm{Semi}}/\Delta t_{\mathrm{Exp,m}}$, in which $\Delta t_{\mathrm{Exp,m}}$ is the maximum time step size determined by the CFL condition of the explicit FDTD method.

7.3 Analysis of a 2-D Polarizer with an InAs Layer

The polarizers to be analyzed are shown in Fig. 7.1. The TE wave is eliminated by interference between the higher-order and lower-order modes, and only the TM wave is guided. The core and substrate are selected as Si [9] and SiO_2 [10], respectively, with relative permittivities of 3.4^2 and 1.94^2. The core width is set to $d = 35\ \mu\text{m}$, and a buffer layer of thickness $t_b = 15\ \mu\text{m}$ is loaded on one side, and InAs of thickness $t_i = 50\ \mu\text{m}$ is loaded on the outside of the buffer. The dielectric constant of the buffer layer ε_b is chosen to be twice the dielectric constant of the core Si, based on the minimum loss condition of TM waves [11]. The polarizer length is set to $l = 94\ \mu\text{m}$. In this subsection, the temperature is set to 300 K and the frequency is set to 1.4 THz. The relative permittivity of InAs in the THz band is expressed by the Drude model [12], which takes in to account the temperature characteristics. Here, each parameter of (7.1) is expressed as [13]

$$\begin{aligned} \varepsilon_\infty &= 15 \\ \omega_p &= 7.19072 \times 10^{12} \\ &\quad + 1.84483 \times 10^{10} \times \exp(0.01276T) \\ \Gamma &= 8.62331 \times 10^{11} - 8.61673 \times 10^{9} T \\ &\quad + 3.48223 \times 10^{7} T^2 - 3.66937 \times 10^{4} T^3 \end{aligned}$$

The size of the computational domain is –200 to 200 μm in the x direction and 0 to 1100 μm in the z direction. Convolutional Perfectly Matched Layer (CPML) [14–16] is used as an absorbing boundary condition outside the computational domain.

The guided mode power is calculated as the product of a term evaluating the correlation between the eigenmode field of the waveguide (ϕ_0) and the propagating field obtained from the FDTD method (ϕ) and the normalized power remaining in the computational domain, using the following formula:

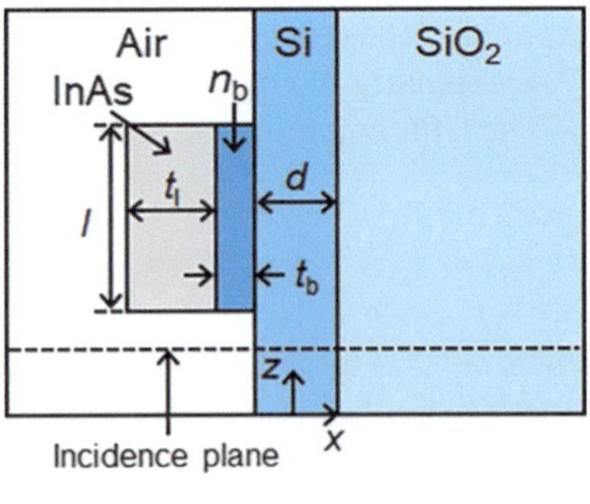

Fig. 7.1 Configuration of a 2-D THz waveguide-type polarizer. Copyright (c) 2022 IEICE from [7]

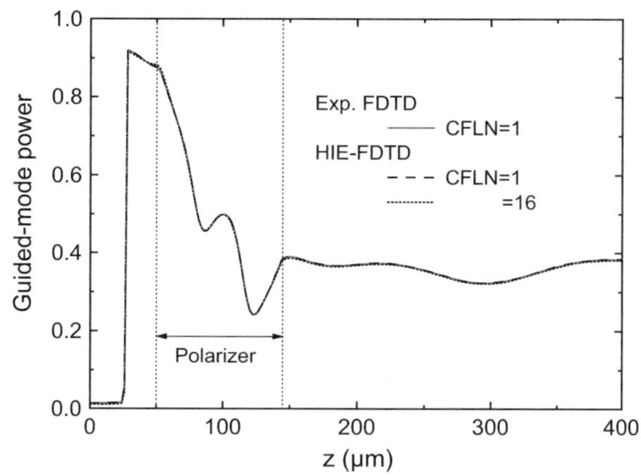

Fig. 7.2 Guided-mode power of the TM mode. Copyright (c) 2022 IEICE from [7]

$$GMP = \frac{|\int \phi_0 \phi^* dx|^2}{\int |\phi_0|^2 dx \int |\phi|^2 dx} \times \frac{\int |\phi|^2 dx}{\int |\phi_0|^2 dx} \tag{7.6}$$

The field components in the above equation are the electric or magnetic field after conversion to a complex field. If the eigenmodes injected into the waveguide propagate without loss, the guied-mode power remains 1. The fundamental mode field in a three-layer slab waveguide is used as the eigenmode field of the waveguide throughout the entire waveguide, from the incident waveguide, polarizer, and output section.

Figure 7.2 shows the guided-mode power as a function of propagation direction, in which the result obtained from the traditional explicit FDTD method is included. In this analysis, the spatial sampling widths are chosen to be $\Delta x = 0.125\ \mu\text{m}$ and $\Delta z = 2.0\ \mu\text{m}$. This results in a maximum CFLN of 16, which corresponds to the aspect ratio of the sampling mesh. It can be seen that the results of the HIE-FDTD method are in excellent agreement with that of the explicit FDTD method, even at a maximum CFLN of 16. In this case, the computation time of the HIE-FDTD method is reduced to 36% of that of the explicit counterpart with a maximum error of 0.81% in the guided-mode power.

The guided-mode power of the TE mode that should be rejected is shown in Fig. 7.3. As in the TM mode, the result of the TE mode from the HIE-FDTD method with CFLN = 16 agrees with that of the explicit counterpart.

Figure 7.4 illustrates the field distributions of the TE and TM modes. As can be seen, for the TE mode, the input fundamental field is transformed into the first-order mode in the polarizer section. Note that the first-order mode cannot propagate through the output waveguide, since the waveguide is a single mode one that rejects higher-order modes. The TM mode power remains in the output waveguide.

The extinction ratio of the TE and TM modes as a function of frequency is shown in Fig. 7.5. As expected from the above results, the ratio from the HIE-FDTD method

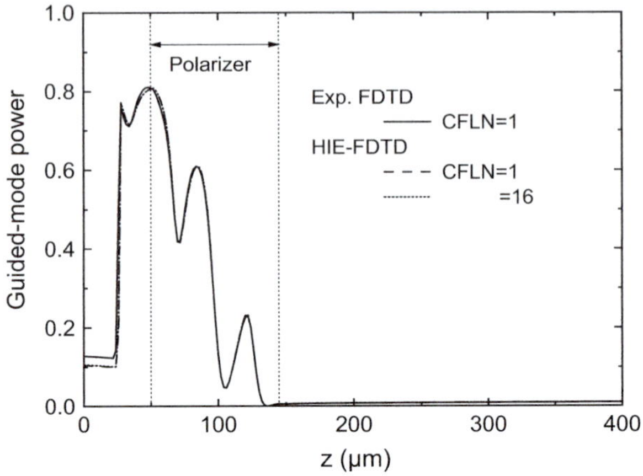

Fig. 7.3 Guided-mode power of the TE mode. Copyright (c) 2022 IEICE from [7]

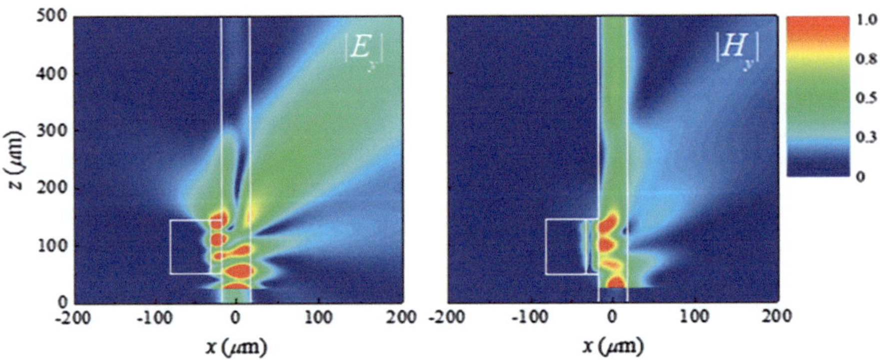

(a) TE mode

(b) TM mode

Fig. 7.4 Propagating field distributions. Copyright (c) 2022 IEICE from [7]

Fig. 7.5 Extinction ratio of the TE and TM modes. Copyright (c) 2022 IEICE from [7]

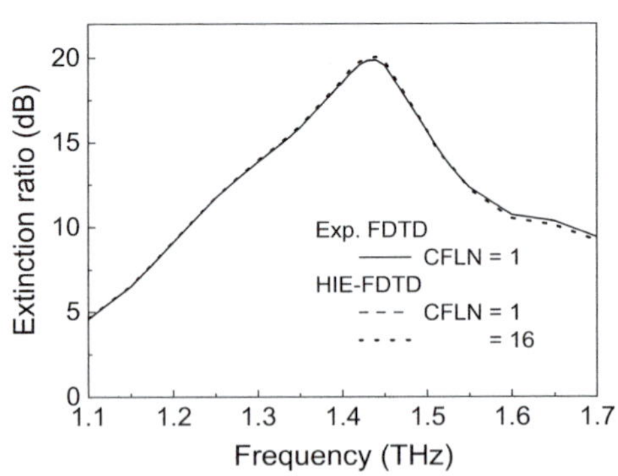

is in good agreement with the explicit result. The polarizer operates with an extinction ratio of more than 15 dB, ranging from 1.33 to 1.53 THz.

7.4 Formulation of the 3-D Frequency-Dependent HIE-FDTD Method

Since the characteristics of the 2-D polarizer have been investigated, we next study a 3-D polarizer showin in Fig. 7.6 (detailed structural parameters will be given later). First, we show the formulation of the 3-D HIE-FDTD method based on the TRC technique. Since a small sampling width is required in the thickness direction of the polarizer, we adopt the implicit method only to the y direction in Fig. 7.6. After some manipulation, we derive the following 3-D HIE-FDTD equations:

$$\begin{aligned} E_x^{n+1} = &\frac{\varepsilon_\infty - \frac{\chi^0}{2}}{\varepsilon_\infty + \frac{\chi^0}{2}} E_x^n + \frac{c\Delta t}{\varepsilon_\infty + \frac{\chi^0}{2}} \frac{\partial}{\partial y}\left(\frac{H_z^{n+1} + H_z^n}{2}\right) \\ &- \frac{c\Delta t}{\varepsilon_\infty + \frac{\chi^0}{2}} \frac{\partial H_y^{n+\frac{1}{2}}}{\partial z} + \frac{1}{\varepsilon_\infty + \frac{\chi^0}{2}} \phi_x^n \end{aligned} \tag{7.7}$$

$$\begin{aligned} E_y^{n+\frac{1}{2}} = &\frac{\varepsilon_\infty - \frac{\chi^0}{2}}{\varepsilon_\infty + \frac{\chi^0}{2}} E_y^{n-\frac{1}{2}} + \frac{c\Delta t}{\varepsilon_\infty + \frac{\chi^0}{2}} \frac{\partial H_x^n}{\partial z} \\ &- \frac{c\Delta t}{\varepsilon_\infty + \frac{\chi^0}{2}} \frac{\partial H_z^n}{\partial x} + \frac{1}{\varepsilon_\infty + \frac{\chi^0}{2}} \phi_y^{n-\frac{1}{2}} \end{aligned} \tag{7.8}$$

$$\begin{aligned} E_z^{n+1} = &\frac{\varepsilon_\infty - \frac{\chi^0}{2}}{\varepsilon_\infty + \frac{\chi^0}{2}} E_z^n - \frac{c\Delta t}{\varepsilon_\infty + \frac{\chi^0}{2}} \frac{\partial}{\partial y}\left(\frac{H_x^{n+1} + H_x^n}{2}\right) \\ &+ \frac{c\Delta t}{\varepsilon_\infty + \frac{\chi^0}{2}} \frac{\partial H_y^{n+\frac{1}{2}}}{\partial x} + \frac{1}{\varepsilon_\infty + \frac{\chi^0}{2}} \phi_z^n \end{aligned} \tag{7.9}$$

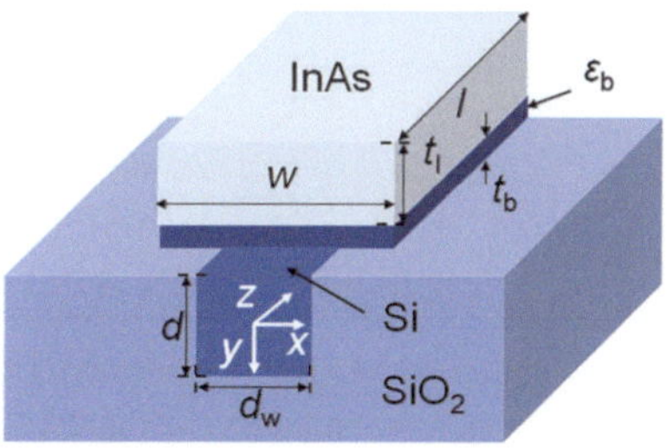

Fig. 7.6 Configuration of a 3-D THz waveguide-type polarizer. Copyright (c) 2024 IEICE from [8]

$$H_x^{n+1} = H_x^n + \frac{c\Delta t}{\mu_r}\frac{\partial E_y^{n+\frac{1}{2}}}{\partial z} - \frac{c\Delta t}{\mu_r}\frac{\partial}{\partial y}\left(\frac{E_z^{n+1} + E_z^n}{2}\right) \tag{7.10}$$

$$H_y^{n+\frac{1}{2}} = H_y^{n-\frac{1}{2}} + \frac{c\Delta t}{\mu_r}\frac{\partial E_z^n}{\partial x} - \frac{c\Delta t}{\mu_r}\frac{\partial E_x^n}{\partial z} \tag{7.11}$$

$$H_z^{n+1} = H_z^n + \frac{c\Delta t}{\mu_r}\frac{\partial}{\partial y}\left(\frac{E_x^{n+1} + E_x^n}{2}\right) - \frac{c\Delta t}{\mu_r}\frac{\partial E_y^{n+\frac{1}{2}}}{\partial x} \tag{7.12}$$

where χ^0, ϕ_ξ^n and $\Delta\chi^0$ are the same as those shown in the 2-D formulation, in which $\xi = x$, y and z. The calculation procedure is as follows: (1) (7.8) and (7.11) are explicitly solved, (2) (7.12) and (7.10) are respectively substituted into (7.7) and (7.9), then resultant tri-diagonal equations are implicitly solved, (3) (7.12) and (7.10) are explicitly solved. Here, we present two equations to be solved implicitly as

$$\begin{aligned}
&-\alpha E^{n+1}_{x,\frac{1}{2},1,0} + (1+2\alpha)E^{n+1}_{x,\frac{1}{2},0,0} - \alpha E^{n+1}_{x,\frac{1}{2},-1,0} \\
&= \beta E^n_{x,\frac{1}{2},0,0} + \alpha\left(E^n_{x,\frac{1}{2},1,0} - 2E^n_{x,\frac{1}{2},0,0} + E^n_{x,\frac{1}{2},-1,0}\right) \\
&\quad - \gamma_x\left(E^{n+\frac{1}{2}}_{y,1,\frac{1}{2},0} - E^{n+\frac{1}{2}}_{y,1,-\frac{1}{2},0} - E^{n+\frac{1}{2}}_{y,0,\frac{1}{2},0} + E^{n+\frac{1}{2}}_{y,0,-\frac{1}{2},0}\right) \\
&\quad + \frac{\zeta}{\Delta y}\left(H^n_{z,\frac{1}{2},\frac{1}{2},0} - H^n_{z,\frac{1}{2},-\frac{1}{2},0}\right) \\
&\quad - \frac{\zeta}{\Delta z}\left(H^{n+\frac{1}{2}}_{y,\frac{1}{2},0,\frac{1}{2}} - H^{n+\frac{1}{2}}_{y,\frac{1}{2},0,-\frac{1}{2}}\right) \\
&\quad + \frac{1}{\varepsilon_\infty + \frac{\chi^0}{2}}\phi^n_{x,\frac{1}{2},0,0}
\end{aligned} \tag{7.13}$$

$$\begin{aligned}
&-\alpha E^{n+1}_{z,0,1,\frac{1}{2}} + (1+2\alpha)E^{n+1}_{z,0,0,\frac{1}{2}} - \alpha E^{n+1}_{z,0,-1,\frac{1}{2}} \\
&= \beta E^n_{z,0,0,\frac{1}{2}} + \alpha\left(E^n_{z,0,1,\frac{1}{2}} - 2E^n_{z,0,0,\frac{1}{2}} + E^n_{z,0,-1,\frac{1}{2}}\right) \\
&\quad - \gamma_z\left(E^{n+\frac{1}{2}}_{y,0,\frac{1}{2},1} - E^{n+\frac{1}{2}}_{y,0,\frac{1}{2},0} - E^{n+\frac{1}{2}}_{y,0,-\frac{1}{2},1} + E^{n+\frac{1}{2}}_{y,0,-\frac{1}{2},0}\right) \\
&\quad + \frac{\zeta}{\Delta x}\left(H^{n+\frac{1}{2}}_{y,\frac{1}{2},0,\frac{1}{2}} - H^{n+\frac{1}{2}}_{y,-\frac{1}{2},0,\frac{1}{2}}\right) \\
&\quad - \frac{\zeta}{\Delta y}\left(H^n_{x,0,\frac{1}{2},\frac{1}{2}} - H^n_{x,0,-\frac{1}{2},\frac{1}{2}}\right) \\
&\quad + \frac{1}{\varepsilon_\infty + \frac{\chi^0}{2}}\phi^n_{z,0,0,\frac{1}{2}}
\end{aligned} \tag{7.14}$$

where

$$\alpha = \frac{c^2 \Delta t^2}{4\left(\varepsilon_\infty + \frac{\chi^0}{2}\right)\Delta y^2}$$

$$\beta = \frac{\varepsilon_\infty - \frac{\chi^0}{2}}{\varepsilon_\infty + \frac{\chi^0}{2}}$$

$$\gamma_x = \frac{c^2 \Delta t^2}{2\mu_r\left(\varepsilon_\infty + \frac{\chi^0}{2}\right)\Delta x \Delta y}, \quad \gamma_z = \frac{c^2 \Delta t^2}{2\mu_r\left(\varepsilon_\infty + \frac{\chi^0}{2}\right)\Delta y \Delta z}$$

$$\zeta = \frac{c\Delta t}{\left(\varepsilon_\infty + \frac{\chi^0}{2}\right)}$$

Note that in the above equations, we omit the directional components x, y and z in the subscript of the field, e.g., $E^{n+1}_{x,1/2,1,0}$ means $E^{n+1}_{x,i+1/2,j+1,k}$. As can be seen, these equations are easily solved with the Thomas algorithm.

The CFL condition of the 3-D HIE-FDTD method is expressed as

$$\Delta t_{\text{Semi}} \leq \frac{n_{\min}}{c\sqrt{(1/\Delta x)^2 + (1/\Delta z)^2}} \tag{7.15}$$

In the following analysis, we use $\Delta x = \Delta z = 2.5\ \mu\text{m}$, $\Delta y = 0.125\ \mu\text{m}$. in which the maximum CFLN is calculated to be 14.

7.5 Analysis of a 3-D Polarizer with an InAs Layer

Figure 7.6 shows the 3-D THz waveguide-type polarizer. The structural parameters are as follows: $d_{\text{w}} = 50\ \mu\text{m}$, $d = 50\ \mu\text{m}$, $w = 200\ \mu\text{m}$, $t_{\text{b}} = 12\ \mu\text{m}$, $t_{\text{I}} = 50\ \mu\text{m}$. First, we present the eigenmode fields of the input Si waveguide. Figure 7.7 illustrates the input TE and TM eignmodes at 1.4 THz. The electric field at the interface between different materials has a field discontinuity due to the boundary condition, while the magnetic field is continuous. This eigenmode field is excited at $z = 50\ \mu\text{m}$, so that it propagates only in the z direction.

Figure 7.8 shows the propagating field distributions at the center of the core for the 3-D polarizer, calculated from the explicit FDTD method and the HIE-FDTD method with a maximum CFLN of 14. It can be seen that even at the maximum CFLN, the propagating fields agree quite well with those of the explicit counterparts for both the TE and TM modes. For the TE mode, the input fundamental mode field is converted to the first mode in the polarizer. The first mode cannot propagate through the single mode waveguide, so that the wavguide acts as a TE mode stop polarizer if the polarizer section is terminated where the first mode excites.

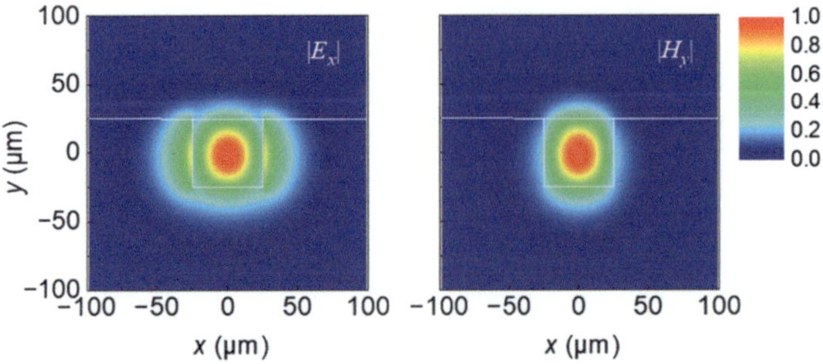

(a) TE mode

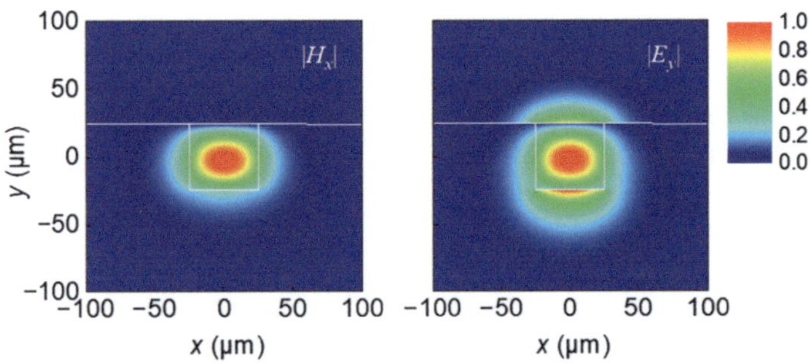

(b) TM mode

Fig. 7.7 Input eigenmode fields. Copyright (c) 2024 IEICE from [8]

Next, we calculate the guided-mode power for the TE and TM modes. The guided-mode power is calculated from the product of the normalized power, which evaluates the remaining power in the waveguide, and the overlap integral, which evaluates the correlation between the input eigemode field and the propagating one. The normalized power and the overlap integral for the full-wave calculations are expressed as

$$NP = \frac{\left|\iint \mathrm{Re}\left[E_x \cdot H_y^*\right] dxdy\right| - \left|\iint \mathrm{Re}\left[E_y \cdot H_x^*\right] dxdy\right|}{\left|\iint \mathrm{Re}\left[E_{x0} \cdot H_{y0}^*\right] dxdy\right| - \left|\iint \mathrm{Re}\left[E_{y0} \cdot H_{x0}^*\right] dxdy\right|} \tag{7.16}$$

$$OV = \frac{\left|\iint E_x \cdot H_{y0}^* - E_y \cdot H_{x0}^* dxdy\right|}{\mathrm{Re}\left[\iint E_x \cdot H_y^* - E_y \cdot H_x^* dxdy\right]} \cdot \frac{\left|\iint E_{x0} \cdot H_y^* - E_{y0} \cdot H_x^* dxdy\right|}{\mathrm{Re}\left[\iint E_{x0} \cdot H_{y0}^* - E_{y0} \cdot H_{x0}^* dxdy\right]} \tag{7.17}$$

where the subscript, x0, y0, represents the reference field selected as the input fundamental eigenmode field, and * represents the complex conjugate.

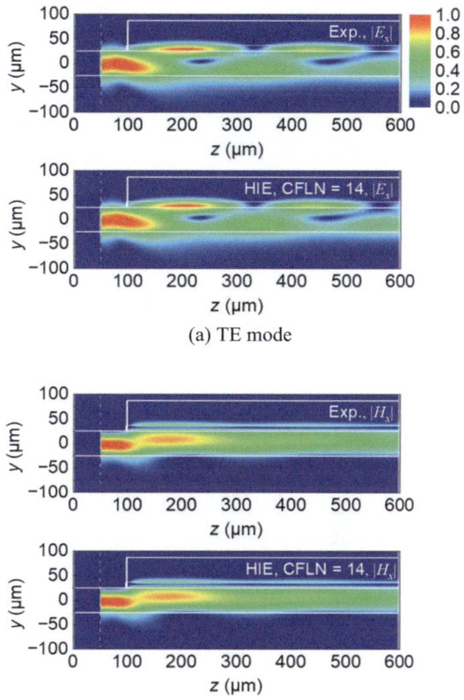

Fig. 7.8 Field distributions at the center of the core. Copyright (c) 2024 IEICE from [8]

Figure 7.9 shows the guided-mode power as a function of propagation distance. It is found that the guided-mode power of the HIE-FDTD method is in almost perfect agreement to that of the explicit FDTD method. For the TE mode, the power oscillates due to the interference between the fundamental and first-order modes. For the TM mode, the power remains in the polarizer section.

From these results, the extinction ratio is calculated from

$$ER = 10 \log_{10} \frac{P_{\mathrm{TM}}}{P_{\mathrm{TE}}} \tag{7.18}$$

where P_{TE} and P_{TM} are the guided-mode power for the TE and TM modes, respectively. The extinction ratio is shown in Fig. 7.10. As expected from Fig. 7.9, the extinction ratio of the HIE-FDTD method is seen to agree well with the explicit counterpart. A extinction ratio of 28 dB is achieved for $z = 266$ μm corresponding to a polarizer length of $l = 126$ μm.

Figure 7.11 illustrates the field distributions when the polarizer is terminated at $z = 266$ μm. For the TE mode, the higher mode is excited in the polarizer section and it radiates in the output waveguide section. The TM mode field propagates through the output waveguide. Note that there is a standing wave in the polarizer section that is not present in Fig. 7.8, due to the reflected field from the end facet of the polarizer.

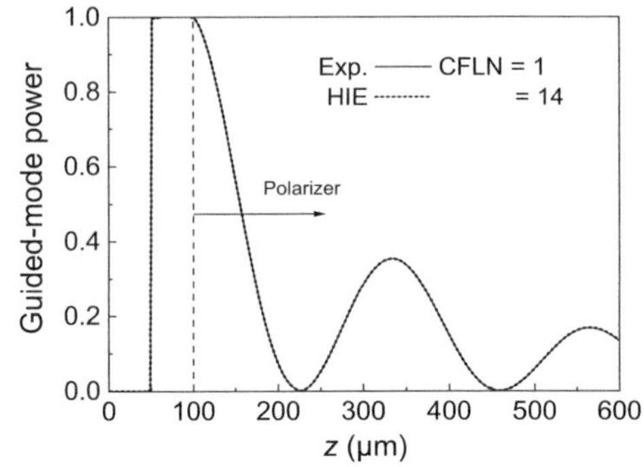

(a) TE mode

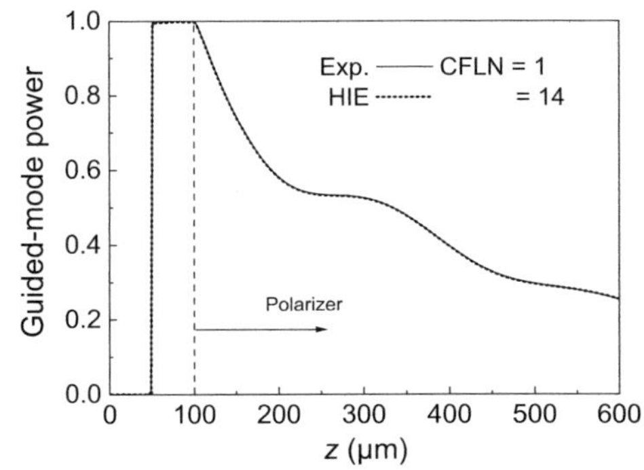

(b) TM mode

Fig. 7.9 Guided-mode power. Copyright (c) 2024 IEICE from [8]

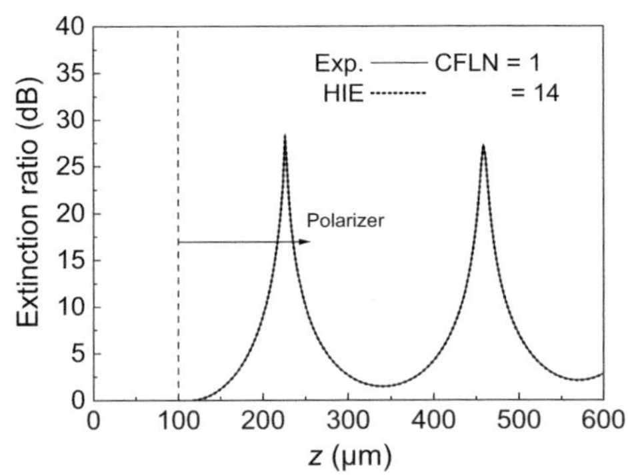

Fig. 7.10 Extinction ratio of the TE and TM modes. Copyright (c) 2024 IEICE from [8]

7.6 Calculation Accuracy of the TRC and RC Techniques

It should be noted that the TRC technique is required for analysis using the HIE-FDTD method with a large CFLN. Figure 7.12 shows the guided-mode power, comparing the TRC and RC results. For the explicit FDTD method, the results of the

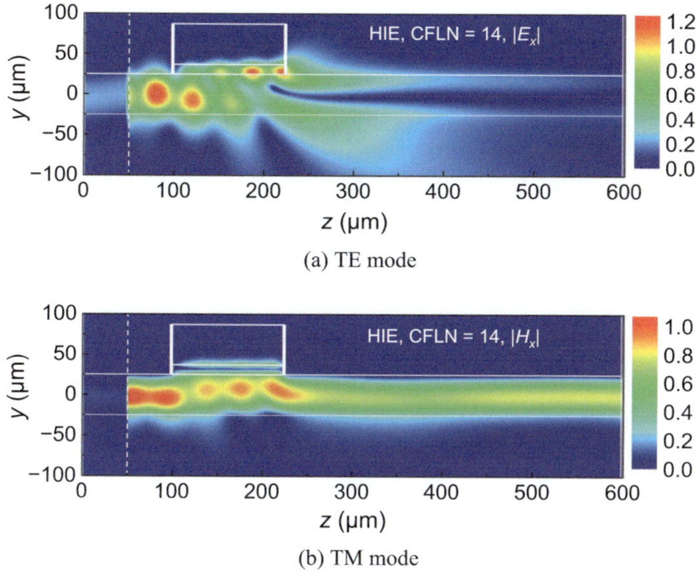

Fig. 7.11 Field distributions for a polarizer length of $l = 126\ \mu\text{m}$. Copyright (c) 2024 IEICE from [8]

TRC and RC techniques are in good agreement for both the TE and TM modes. However, for the HIE-FDTD method, the result of the RC technique at CFLN = 14 deviates from that of the explicit result. Fortunately, even at CFLN = 14, the result of the HIE-FDTD method using the TRC technique agree quite well with the explicit results. Therefore, when we use a large CFLN for the HIE-FDTD method, we need to employ the TRC technique.

7.7 Comparison Between the 3-D and 2-D Analysis

Here, we point out that a 3-D analysis is required for the analysis of practical devices, paying attention to the comparison between 3-D and 2-D results. We compare the 3-D result discussed in 8.5 with that where the 2-D analysis is performed on the y-z cross-section of the core center of the 3-D polarizer. Note that in 8.3 the core width was chosen to be $d = 35\ \mu\text{m}$, while here it is chosen to be $d = 50\ \mu\text{m}$, since this choice leads to an increase in the TM mode power with a consequent increase in the extinction ratio for the 3-D polarizer.

Figure 7.13 shows the guided-mode power of the TE and TM modes comparing the results of the 3D and 2D analyses. The results of the 2D analysis do not match the results of the 3D analysis, because the 2D analysis cannot account for the field spread in the transverse (x) direction, resulting in incorrect evaluation of the waveguide

Fig. 7.12 Guided-mode power of the TRC and RC techniques. Copyright (c) 2024 IEICE from [8]

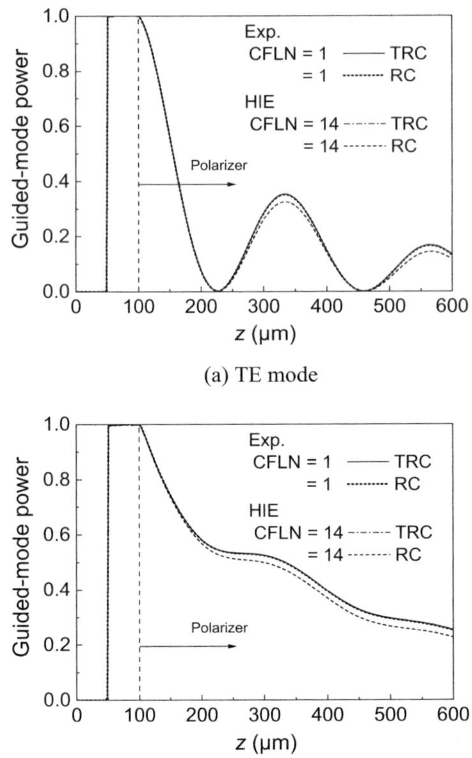

(a) TE mode

(b) TM mode

mode characteristics of the 3D structure. Therefore, 3D analysis is essential for actual device design.

7.8 A 3-D Polarizer with InAs Sidewalls

The polarizers discussed above have exploited the cutoff property of the higher-order TE mode that results from the interference between the fundamental and higher-order modes. As a result, the extinction ratio characteristics are quite sensitive to the polarizer length, as shown in Fig. 7.10. In this section, we propose another polarizer structure in which InAs sidewalls together with dielectric buffer layers are placed on both sides of the core of the waveguide. Only the TE mode is attenuated by the excitation of the lossy surface plasmon, while the TM mode propagates through the waveguide. We analyze the 3-D polarizer with the InAs sidewalls using the HIE-FDTD developed in 8.4. Here we introduce the implicit scheme only in the x direction of the 3-D HIE-FDTD method.

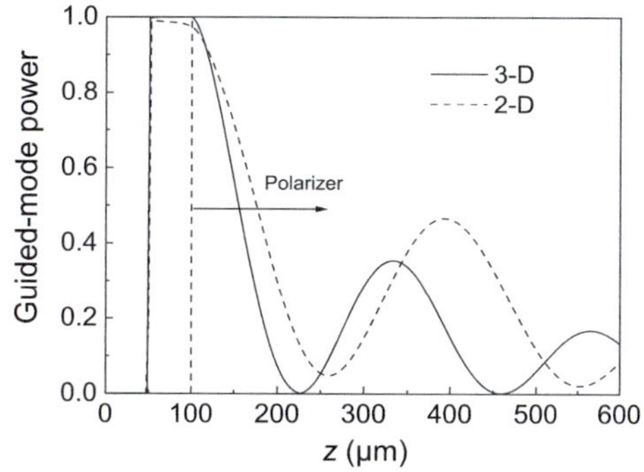

(a) TE mode

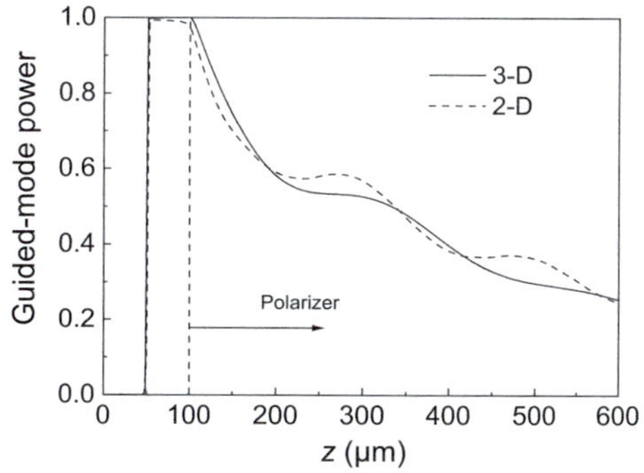

(b) TM mode

Fig. 7.13 Guided-mode power of the 3-D and 2-D analysis. Copyright (c) 2024 IEICE from [8]

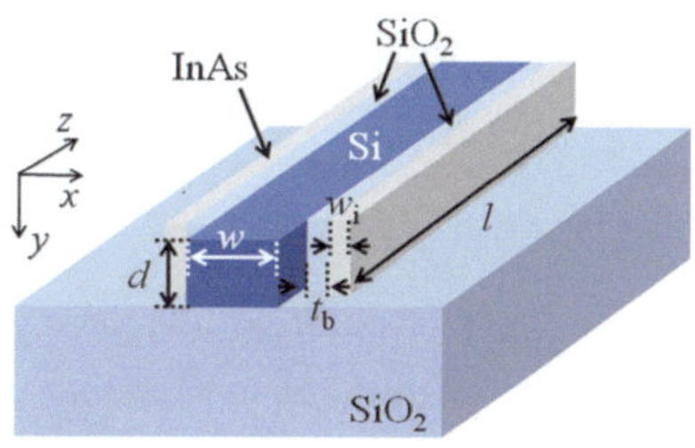

Fig. 7.14 Configuration of a 3-D THz polarizer with InAs sidewalls

Figure 7.14 shows the configuration of the polarizer. The core width and height are $w = d = 50$ μm. SiO_2 buffer layers and InAs sidewalls are placed on both sides of the core, in which $t_b = 13.5$ μm and $w_i = 5$ μm are used. The length of the polarizer section is defined by l. The temperature is set to $T = 300$ K and the designed frequency is 1.4 THz. The spatial sampling widths are $\Delta y = \Delta z = 2.5$ μm, $\Delta x = 0.125$ μm, resulting in a maximum CFLN of 14.

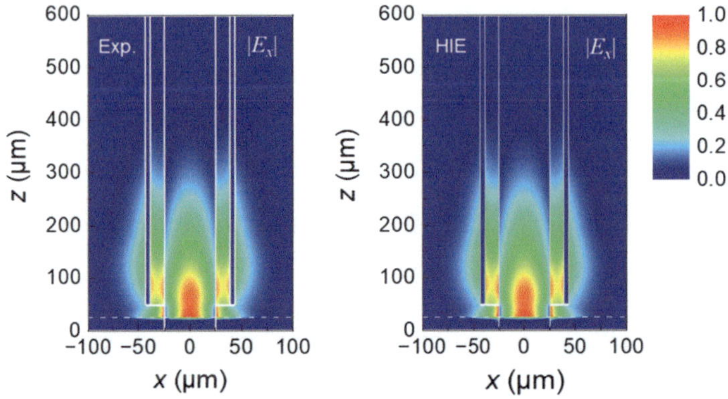

(a) TE mode

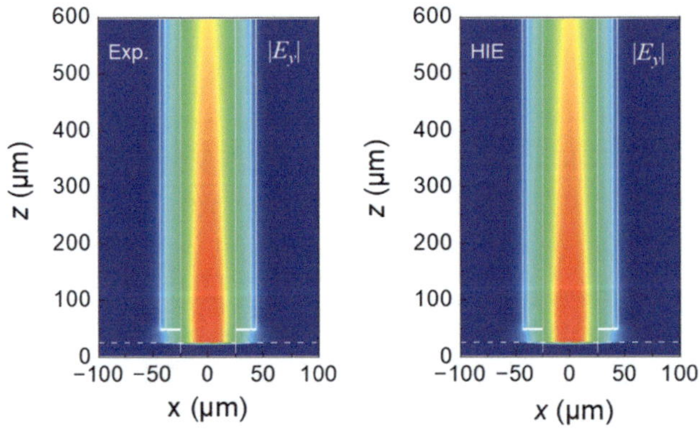

(b) TM mode

Fig. 7.15 Field distributions for the polarizer with InAs sidewalls

Figure 7.15a and b show the field distributions for the TE and TM modes, respectively. The left figures are for the results obtained from the explicit FDTD method and the right figures are for the results obtained from the HIE-FDTD method with CFLN = 14. It can be seen that the HIE results are in good agreement with the explicit counterparts. In this analysis, it is worth mentioning that the computation time of the HIE-FDTD method with CFLN = 14 is reduced to 25% of the explicit counterpart. For the TE mode, the power is attenuated due to the excitation of lossy surface plasmon on the IaAs surface. In contrast, the TM mode maintains its power through the polarizer.

Fig. 7.16 Guided-mode power

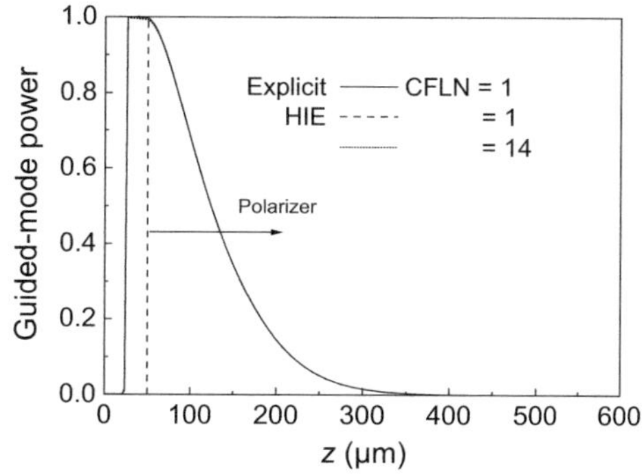

(a) TE mode

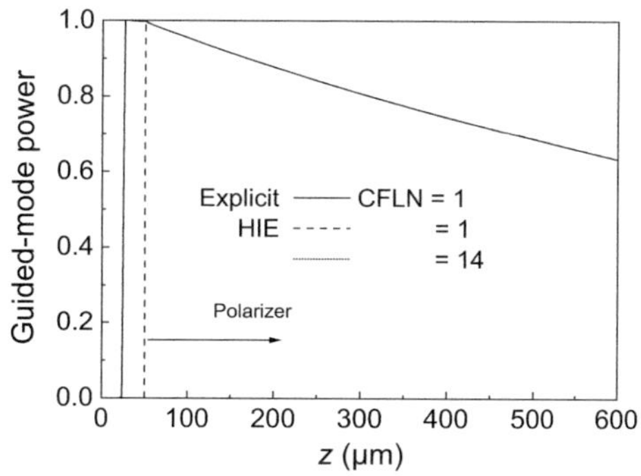

(b) TM mode

Fig. 7.17 Extinction ratio of the TE and TM modes

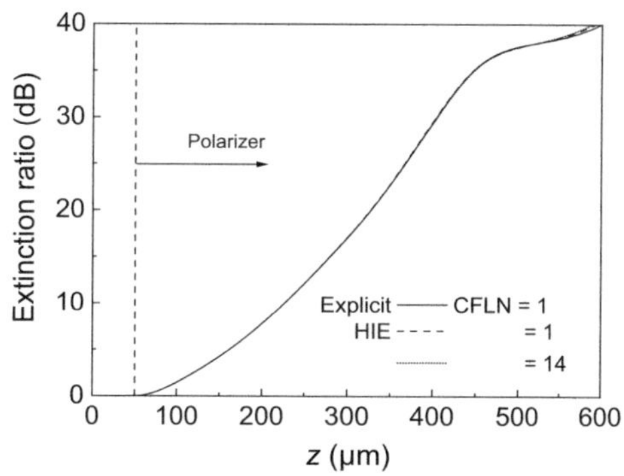

The guided-mode power is shown in Fig. 7.16a and b for the TE and TM modes, respectively. As can be seen, the results of the HIE-FDTD method are superimposed on the explicit results for both modes. From these results, the extinction ratio is calculated in Fig. 7.17, in which the HIE result with CFLN = 14 agrees well with the explicit result. The polarizer shows an extinction ratio of more than 20 dB for a polarizer length of more than $l = 200$ μm. Note that the extinction ratio increases monotonically with the polarizer length, which is effective for a practical device.

7.9 A 3-D Wavegude with Air Buffers and InAs Sidewalls

Our final investigation is given to a 3-D waveguide with air buffers and InAs sidewalls, shown in Fig. 7.18. The materials used for the waveguide are the same as those in Fig. 7.14. The core dimension is slightly modified to $w = 50$ μm and $d = 60$ μm. InAs sidewalls of w_i with air buffers are placed on both sides of the core, in which $t_b = 12.5$ μm and $w_i = 5$ μm are used. The temperature is set to $T = 300$ K and the designed frequency is 1.3 THz. The spatial sampling widths are $\Delta y = \Delta z = 2.5$ μm, $\Delta x = 0.125$ μm, resulting in a maximum CFLN of 14.

Figure 7.19 shows the field distributions for the 3-D waveguide with the buffers being filled with air. It can be seen that the fields calculated from the HIE-FDTD method with CFLN=14 agree quite well with those from the explicit counterpart. It is worth mentioning that both TE and TM waves propagate through the waveguide. The guided-mode power is shown in Fig. 7.20, in which all the results are superimposed. Compared with the TM mode, the power of the TE mode is attenuated since the amplitude of the field outside the core is large, as shown in Fig. 7.19, resulting in a large propagation loss.

Next, we fill the buffers with water and expect the waveguide to act as a polarizer. Figure 7.21 shows the field distributions when the buffers are filled with water. As can be seen, the TE wave rapidly attenuates due to surface plasmon excitation, while the TM wave propagates through the waveguide. Figure 7.22 shows the guided-mode power, which indicates the performance of the polarizer. Using both the TE and TM guided-mode powers, the extinction ratio is calculated in Fig. 7.23. More than $z = 300$ μm, an extinction ratio of 20 dB can be achieved, and it increases with the propagation distance. As a result, the waveguide acts as a simple waveguide with air buffers and as a polarizer when the buffers are filled with water.

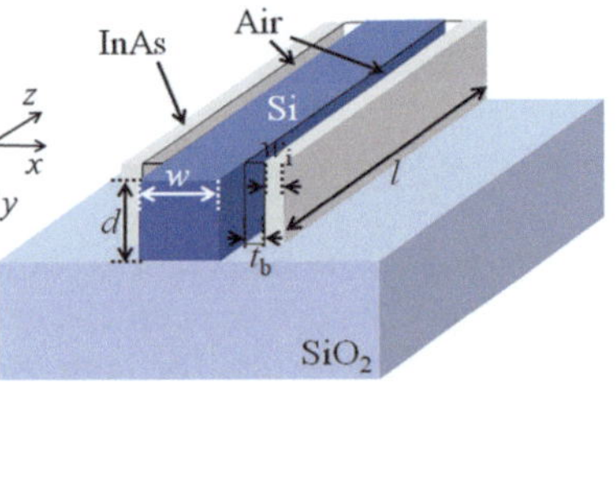

Fig. 7.18 Configuration of a 3-D waveguide with air buffers and InAs sidewalls

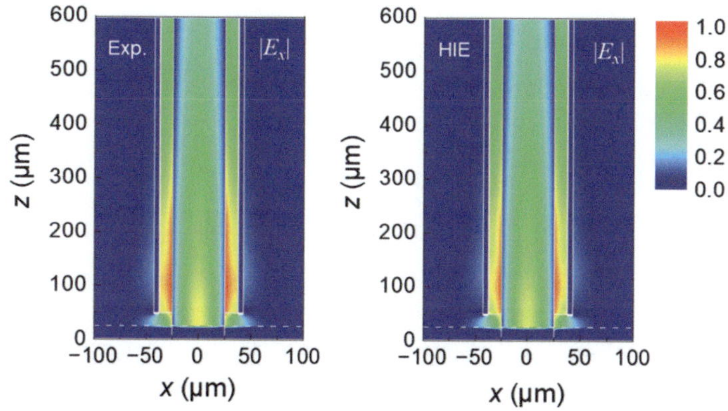

(a) TE mode

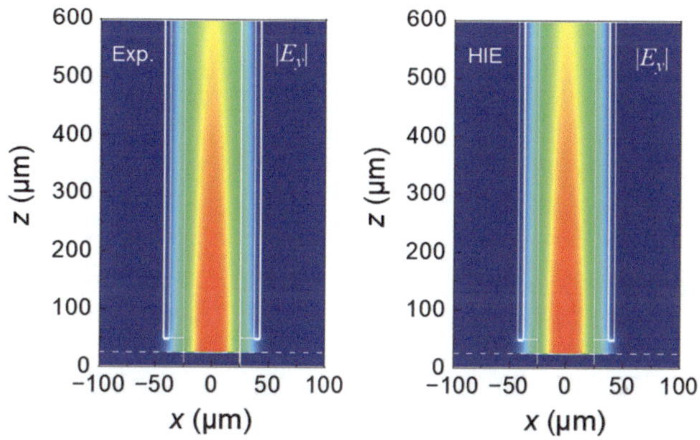

(b) TM mode

Fig. 7.19 Field distributions for the 3-D waveguide with the buffers being filled with air

7.10 Concluding Remarks

A 2-D HIE-FDTD method using the TRC technique has been newly developed to analyze waveguide polarizers with TM-pass/TE-stop characteristics, which provide the input TM wave for the SPR sensor. The guided-mode power obtained from the HIE-FDTD method at the maximum CFLN agrees quite well with that from the explicit FDTD method, in which the maximum error of the guided mode power is 0.81%, and the computation time can be reduced to about 36%. A 3-D HIE-FDTD method has also been developed for the analysis of practical polarizers. The computation time can be reduced to about 22% of that of the traditional explicit FDTD method, while maintaining the calculation accuracy. An extinction ratio of 28 dB is achieved for a polarization length of $l = 125$ μm. The high accuracy of the

Fig. 7.20 Guided-mode power

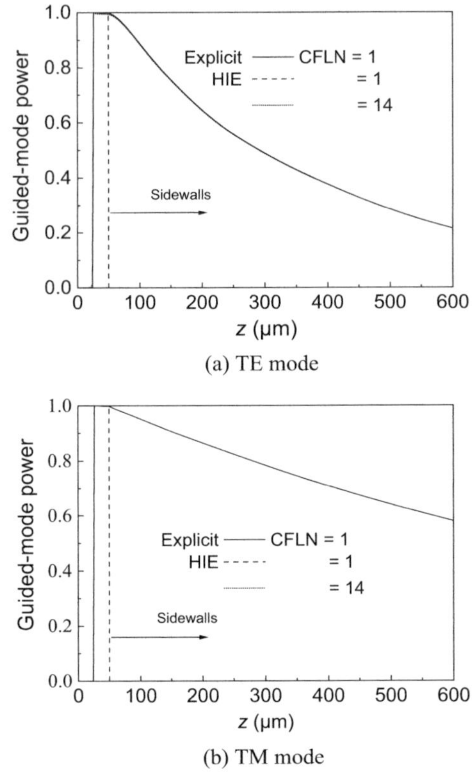

TRC technique is required for analysis using the HIE-FDTD method with a large CFLN. The necessity of the 3D analysis is also discussed for the study of practical polarizers.

We have finally studied another polarizer with InAs sidewalls placed outside the dielectric buffer layers on both sides of the core. For the presented polarizer, the TE waves generate surface plasmons and become lossy, so the field is attenuated and only TM modes are guided. It follows that the polarizer is not sensitive to the device length. We have also considered the case where the dielectric buffer layers are chosen to be hollow. When the buffers are hollow, the waveguide results in TE and TM mode transmission. In contrast, when water is injected into the buffer, the waveguide acts as a TE stop/TM pass polarizer.

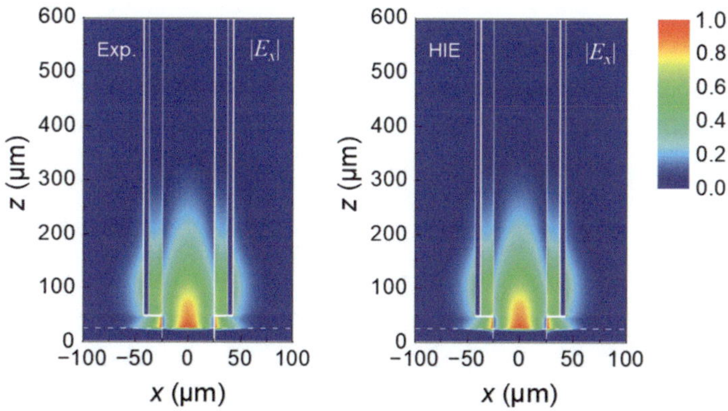

(a) TE mode

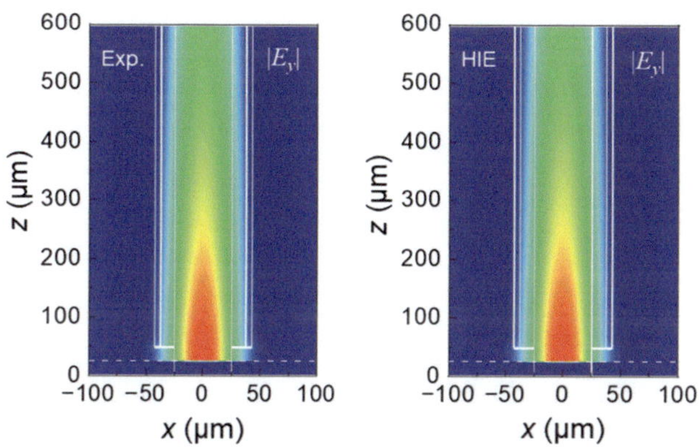

(b) TM mode

Fig. 7.21 Field distributions for the 3-D waveguide with the buffers being filled with water

Fig. 7.22 Guided-mode power

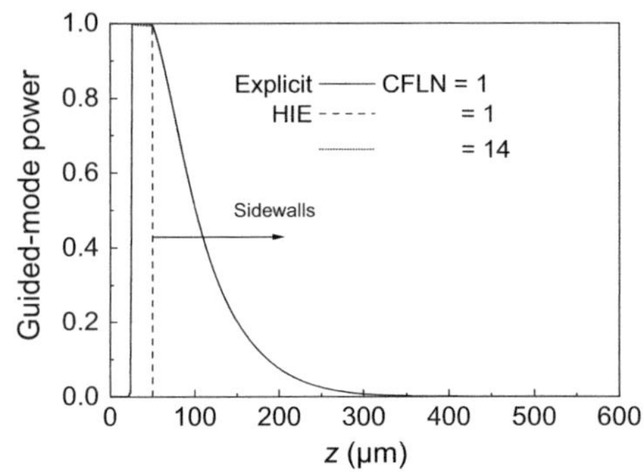

(a) TE mode

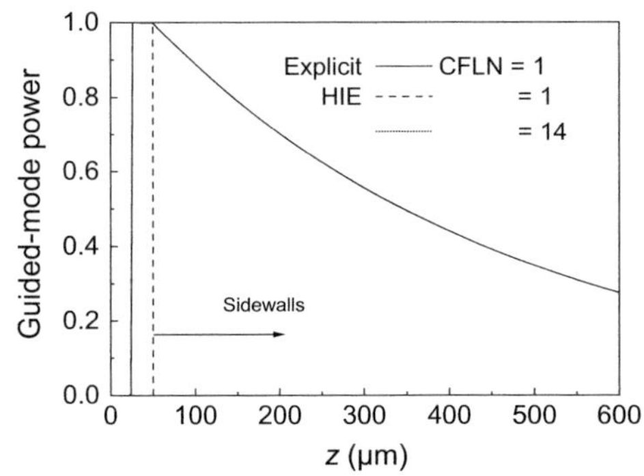

(b) TM mode

Fig. 7.23 Extinction ratio of the TE and TM modes

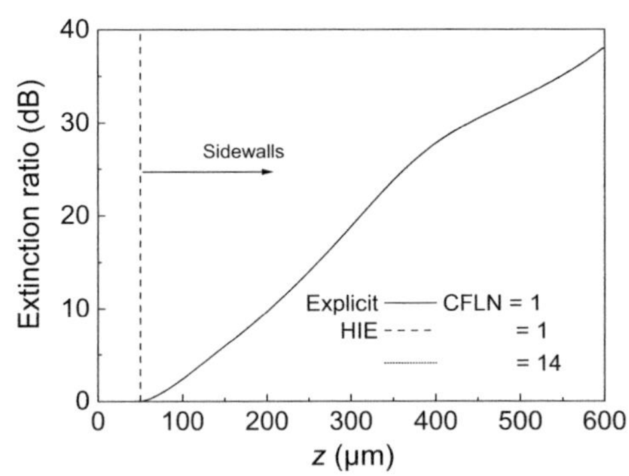

References

1. J. Shibayama, K. Shimizu, J. Yamauchi, H. Nakano, Surface plasmon resonance waveguide sensor in the terahertz regime. J. Light. Technol. **34**(10), 2518–2525 (2016)
2. B. Huang, G. Wang, Y. Jiang, W. Wang, A hybrid implicit-explicit FDTD scheme with weakly conditional stability. Microw. Opt. Technol. Lett. **39**(2), 97–101 (2003)
3. J. Chen, J. Wang, A 3D hybrid implicit-explicit FDTD scheme with weakly conditional stability. Microw. Opt. Technol. Lett. **48**(11), 2291–2294 (2006)
4. K. Fujita, A partially implicit FDTD method for the wideband analysis of spoof localized surface plasmons. IEEE Photonics Technol. Lett. **27**(10), 1124–1127 (2015)
5. H. Muraoka, Y. Inoue, T. Sekine, H. Asai, A hybrid implicit-explicit and conformal (HIE/C) FDTD method for efficient electromagnetic simulation of nonorthogonally aligned thin structures. IEEE Trans. Electromagn. Compat. **57**(3), 505–512 (2015)
6. J. Shibayama, T. Hara, M. Ito, J. Yamauchi, H. Nakano, Improved cylindrical HIE-FDTD method and its application to a metal disc-type surface wave splitter. IEEE Microw. Wirel. Compon. Lett. **29**(3), 177–179 (2019)
7. J. Shibayama, S. Gomi, J. Yamauchi, H. Nakano, Analysis of a TM-pass/TE-stop terahertz waveguide polarizer using implicit and semi-Implicit FDTD methods. IEICE Trans. Electron. **J105-C**(5), 185–192 (2022). (Japanese Edition)
8. H. Miyao, J. Shibayama, J. Yamauchi, H. Nakano, Analysis of a TM-pass terahertz waveguide polarizer using the three-dimensional semi-implicit FDTD method. IEICE Trans. Electron. **J107-C**(5), 200–208 (2024). (Japanese Edition)
9. D. Grischkowsky, S. Keiding, M. van Exter, C. Fattinger, Far-infrared time-domain spectroscopy with terahertz beams of dielectrics and semiconductors. J. Opt. Soc. Am. B **7**(10), 2006–2015 (1990)
10. Y.H. Avetisyan, H.S. Hakobyan, V.R. Tadevosyan, Analysis of plasmon terahertz waveguide formed by a dielectric ridge on semiconductor surface. J. Contemp. Phys. **46**(5), 361–367 (2011)
11. M. Miyagi, A. Hongo, S. Kawakami, Transmission characteristics of dielectric-coated metallic waveguide for infrared transmission: slab waveguide model. IEEE J. Quantum Electron. **QE-19**(2), 136–145 (1983)
12. J.W. Han, M.S. Kim, M.S. Song, B.Y. Kang, B.K. Cho, J.S. Lee, Application of emitter-sample hybrid terahertz time-domain spectoscopy to investigate temperature-dependent optical constants of doped InAs. Appl. Opt. **56**(9), 2529–2534 (2017)
13. K. Yoshihara, J. Shibayama, J. Yamauchi, H. Nakano, Evaluation of the characteristics of a terahertz SPR waveguide sensor with an analyte container. IEICE Tech. Rep. **118**(209), EST2018-49, 35–39 (2018)
14. J.A. Roden, S.D. Gedney, Convolution PML (CPML): an efficient FDTD implementation of the CFS-PML for arbitrary media. Microw. Opt. Technol. Lett. **27**(5), 334–339 (2000)
15. M. Kuzuoglu, R. Mittra, Frequency dependence of the constitutive parameters of causal perfectly matched anisotropic absorbers. IEEE Microw. Guid. Wave Lett. **6**(12), 447–449 (1996)
16. Z. Liu, Y. Chen, X. Sun, Y. Liu, Implementation of CFS-PML for HIE-FDTD method. IEEE Antennas Wirel. Propag. Lett. **11**, 381–384 (2012)

Chapter 8
Grating Consisting of InSb-Coated Dielectric Cylinders and InSb Spheres at Terahertz Frequencies

Abstract This chapter presents the analysis of a grating consisting of periodic arrays of InSb-coated dielectric cylinders on a substrate at THz frequencies using the two-dimensional frequency-dependent finite-difference time-domain method based on the trapezoidal recursive convolution technique. The transmission characteristics of an infinite periodic array are studied not only at normal incidence but also at oblique incidence. It is shown that the incident field is coupled to the substrate due to the guided-mode resonance (GMR), indicating the practical application of a grating coupler. For the sensor application, the frequency shift of the transmission dip is investigated with respect to the variation of the background refractive index. It is found that the shift of the dip involving the surface plasmon resonance is almost ten times larger than that of the dip only from the GMR. We also analyze a finite periodic array of the cylinders. The field radiation from the array is discussed, when the field propagates through the substrate. It is shown that the direction of the radiation can be controlled with the frequency of the propagating field. Final investigation is given to the three-dimensional analysis of InSb spheres array on a dielectric substrate. It is shown that the propagation direction depends on the frequency at which the field is coupled to the substrate. An application to the sensor is also discussed with attention to the behavior of the coupled field.

8.1 Introduction

Optical waveguides with grating structures have been widely used for many applications, such as grating couplers [1–5], filters [6], mode converters [7], polarization splitters [8] and so on. In general, the grating structures have been made of dielectric materials. On the other hand, a grating structure with metal materials has also been investigated, in which the characteristics of a surface plasmon resonance (SPR) are exploited [9]. In [10–12], the scattering properties have been theoretically studied for gratings of metal-coated circular nanocylinders on a dielectric substrate. This work has presented the transmission dip from the guided-mode resonance (GMR) and that involving the SPR.

J. Shibayama, *FDTD Analysis of Plasmonic Devices*, Springer Series in Optical Sciences 254, https://doi.org/10.1007/978-981-96-8130-3_8

At terahertz (THz) frequencies, a metal/dielectric interface cannot support the SPR due to negligible penetration of the electromagnetic fields into the metal. One way to create surface waves with metal is to use an array of holes or gratings on the metal, each much smaller than the wavelength. This results in surface modes on the metal that mimic surface plasmons, called spoof surface plasmon polaritons [13–16]. In contrast, a semiconductor material such as indium antimonide (InSb) exhibits a negative permittivity at THz frequencies, which supports the SPR along the InSb/dielectric interface [17]. Using InSb devices, a number of plasmonic devices have recently been investigated, such as waveguides [18], filters [19], lenses [20] and sensors [21–23].

The first purpose of this section is to investigate the fundamental properties of a grating consisting of InSb-coated dielectric cylinders on a substrate at THz frequencies. The structure discussed here is similar to that found in [13–16] at optical frequencies. However, here we use InSb to excite the SPR at THz frequencies. In [24], the transmission characteristics of the infinite array of the cylinders have been shown only at normal incidence. In this chapter, we discuss the transmission characteristics not only at normal incidence but also at oblique incidence [25]. In addition, a finite periodic array of the cylinders is studied to show the field radiation from the grating of the cylinders.

First, the transmission characteristics at normal and oblique incidence are studied. At the transmission dips, it is shown that the incident field is coupled to the substrate due to the GMR. This leads to an application in a grating coupler where a wave with a specific THz frequency can be extracted from the substrate. We also investigate the effect of the background refractive index on the transmission dip for the sensor application. It is found that the frequency shift of the dip involving the SPR is about ten times larger than that of the dip involving only the GMR. Finally, a finite array of cylinders from which the field radiates into the background space as it propagates through the substrate is studied. It is shown that the radiation direction can be controlled with the frequency of the propagating field.

The second purpose of this section is to investigate a periodic array of InSb spheres on a substrate using the three-dimensional (3-D) FDTD method [26]. It is shown that he incident field is coupled to the eigenmode in the substrate due to GMR. It is also found that the propagation direction of the coupled field depends on the frequency at which the field is coupled to the substrate. This behavior is not observed in the two-dimensional (2-D) problem [24], since a grating structure cannot be modeled in a particular direction and consequently the GMR does not occur in this direction. The effect of the refractive index change of the surrounding and background space on the transmission characteristic is also studied for sensor applications. It is shown that the shift of the transmission dip becomes significant for the case where the localized SPP field widely extends into the background space.

8.2 Analysis of an Infinite Array of Cylinders

The configuration of a 2-D grating consisting of InSb-coated dielectric cylinder arrays on a substrate is shown in Fig. 8.1 (a similar cylinder configuration has been investigated for cloaking [27]). The dispersive characteristics of the InSb is given in Sect. 5.2. Only the unit cell is shown, and its period is chosen to be $\Lambda = 168$ μm, which mainly determines the frequency of the GMR. The material of the dielectric cylinder and the substrate is SiO_2, whose refractive index is $n_s = 1.92$. A relatively small loss of SiO_2 is ignored in this work. The dielectric cylinder is covered with InSb, where $r_1 = 28$ and $r_2 = 14$ μm are chosen. The refractive index of the background is defined by n_B. The thickness of the substrate is set to $t = 140$ μm. These structural parameters are chosen so that the GMR is around 1 THz. Note that this frequency range corresponds to that obtained from photoconductive antennas [28, 29]. To analyze an infinite array in the x direction, we use the FDTD method with the periodic boundary condition, in which the trapezoidal recursive convolution technique is adopted to take into account the Drude model of the InSb permittivity. The E_x (H_y) polarized plane wave is excited along the plane of incidence in the $+z$ direction. The transmission characteristics are studied along the observation plane. Since we are dealing with a two-dimensional problem, the structure in the vertical (y) direction is constant, i.e. $\partial/\partial y = 0$.

8.2.1 *Transmission Characteristics at Normal Incidence*

Figure 8.2 shows the transmission characteristics at normal incidence ($\theta = 0°$) for $T = 300$ K and $n_B = 1.0$. Figure 8.2a shows the results for different InSb layer thicknesses while keeping r_2, and Fig. 8.2b shows the results for different rod radii while keeping the ratio of r_1 and r_2. Note that three transmission dips are observed

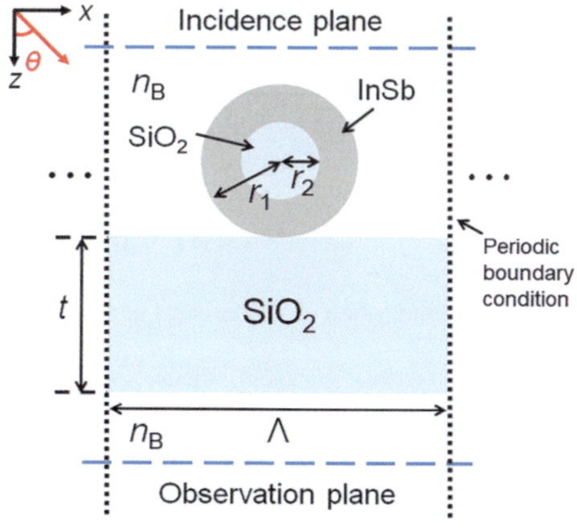

Fig. 8.1 Configuration of a grating consisting of InSb-coated dielectric cylinders on a substrate. Copyright © 2020 IEICE from [25]

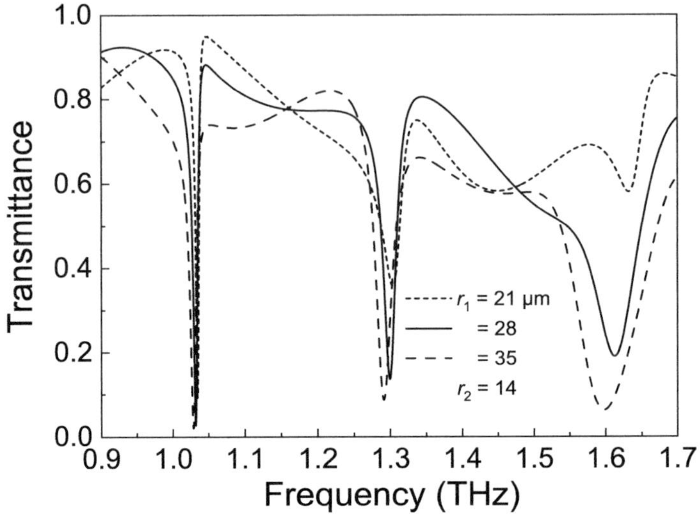

(a) Various thicknesses of the InSb layer

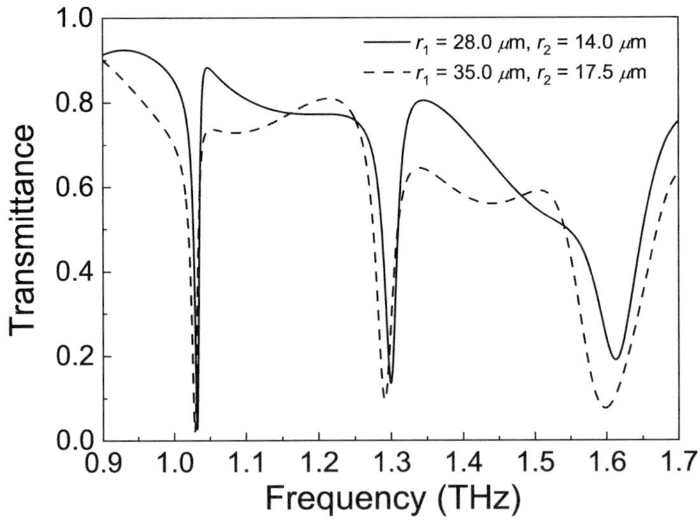

(b) Various radii of the rod

Fig. 8.2 Transmission characteristics at normal incidence. Copyright © 2020 IEICE from [25]

for each data. In Fig. 8.2a, for $r_1 = 21$ and $r_2 = 14$ μm, it can be seen that the transmission dips around 1.3 and 1.63 THz are weaker than the other two. This is because the SPR is not effectively excited due to a thin InSb layer. The dip around

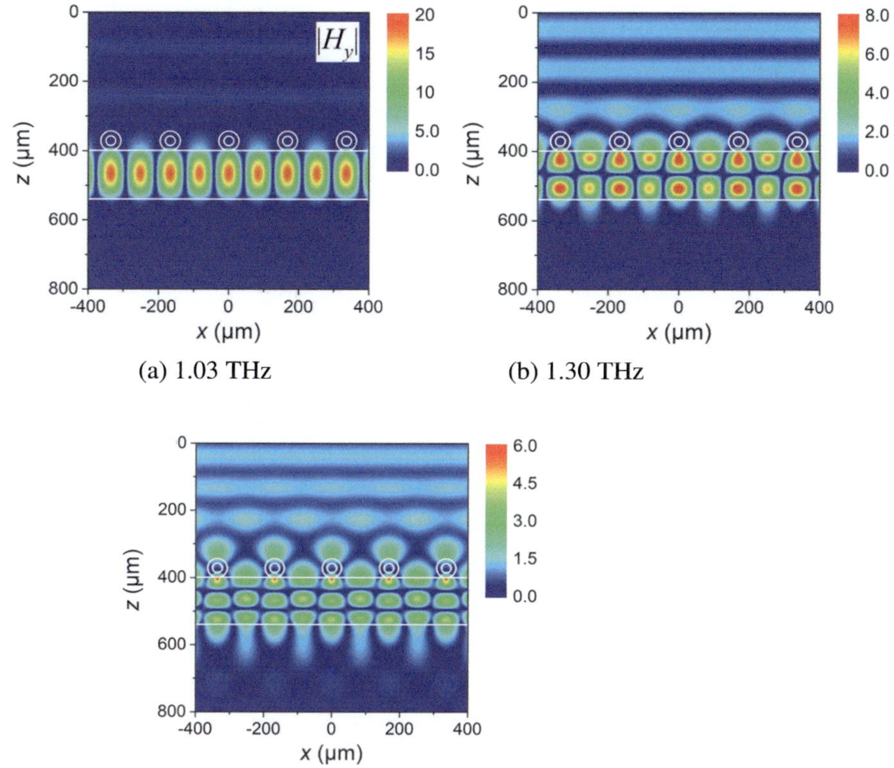

(a) 1.03 THz

(b) 1.30 THz

(c) 1.61 THz

Fig. 8.3 Steady-state fields at the transmission dips at normal incidence for $r_1 = 28$ and $r_2 = 14$ μm. Copyright © 2020 IEICE from [25]

1.03 THz is not so much affected by the InSb thickness because the SPR is not excited around this frequency. In Fig. 8.2b, the result for $r_1 = 35$ and $r_2 = 17.5$ μm is quite similar to that for $r_1 = 35$ and $r_2 = 14$ μm in Fig. 8.2a. This is due to the fact that the outer radius of the rod is the same for a sufficiently thick InSb layer.

Figure 8.3 shows the steady-state fields (H_y) at the three transmission dips in Fig. 8.2. For example, we show the case for $r_1 = 28$ and $r_2 = 14$ μm and observe the dips at 1.03, 1.30, and 1.61 THz. The field is normalized to the maximum amplitude of the incident field. Note that the incident field is coupled to the eigenmode of the substrate as a GMR, provided that the following phase matching condition is satisfied:

$$k \sin \theta = \beta_0 + q \frac{2\pi}{\Lambda} \tag{8.1}$$

where k is the wavenumber, θ is the angle of incidence, β_0 is the value close to the effective index of the mode field in the substrate, q is the diffraction order, and Λ

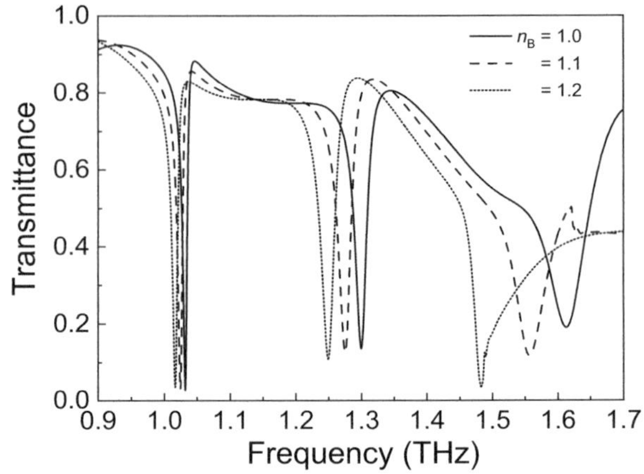

Fig. 8.4 Transmission characteristics for various n_B. Copyright © 2020 IEICE from [25]

is the grating period. At 1.03 THz, the incident field is coupled to the fundamental mode of the substrate, as shown in Fig. 8.3a. In this case, the field propagates in the $\pm x$ directions with $q = \mp 1$ in (8.1), resulting in the standing wave. Similarly, at 1.30 and 1.61 THz, the field is coupled to the first- and second-order modes, respectively. Note that at 1.61 THz, that the field stems from the SPR can be found around the top and bottom of the cylinders. Since the SPR induces an absorption of the field, the response in Fig. 8.2 is more gradual than the other two.

Here we examine the effect of the background refractive index n_B on the transmission dips. Unless otherwise noted, we use the cylinder for $r_1 = 28$ and $r_2 = 14\ \mu$m. Figure 8.4 shows the transmission characteristics for different n_B. It can be seen that the shift of the dip at higher frequencies is much larger than that at lower frequencies. To quantify the frequency shift, we calculate the sensitivity defined by $\delta f/\delta n$, where δf is the frequency shift and δn is that of the refractive index. The sensitivity is calculated to be 7.65×10^{-2}, 2.53×10^{-1}, and 6.51×10^{-1} THz/refractive index unit (RIU) at about 1.0, 1.25, and 1.55 THz, respectively. In particular, the shift at about 1.55 THz is significant. This is because the SPR excited around the cylinders is sensitive to the background refractive index. In addition, the evanescent field extends into the background space as shown in Fig. 8.3c, which has a high ability to detect the refractive index change. This property can be used as a sensor for gas and liquid analysis.

We also observe in Fig. 8.4 that the transmission at about 1.55 THz decreases as n_B is increased. This is due to the fact that the absorption of the SPR increases as the refractive index of the surrounding material increases. Although not shown, the dip at higher frequencies disappears for n_B greater than 1.4, since the excitation condition is no longer satisfied. The dip at about 1.25 THz survives up to $n_B = 1.5$ and the one at about 1.0 THz survives up to $n_B = 1.9$.

8.2.2 Transmission Characteristics at Oblique Incidence

Next we consider the case of oblique incidence. As can be seen in Fig. 8.5, there are several transmission dips for $\theta = 10°$ and $20°$. Here we focus on the result for $\theta = 20°$. Figure 8.6 shows the steady-state fields at the transmission dips in Fig. 8.5. The incident field is coupled to the fundamental mode of the substrate at 0.89 and 1.23 THz, as can be seen in Fig. 8.6a and c. It is interesting to note that the coupled fields propagate in opposite directions at these frequencies. That is, at 0.89 THz, (8.1) is satisfied for a negative β_0 with $q = 1$, so the field propagates in the $-x$ direction. At 1.23 THz, on the other hand, (8.1) is satisfied for a positive β_0 with $q = -1$, so the field propagates in the $+x$ direction. This behavior is confirmed by the Poynting vector calculation. Figure 8.7 shows the results of the Poynting vector. At 0.89 and 1.23 THz, the fields are seen to propagate in the $-x$ and $+x$ directions, respectively.

Note that for oblique incidence, the range that satisfies (8.1) is extended, since the left-hand side of (8.1) is not zero. In the above case, the condition for coupling to the fundamental mode is also satisfied at higher frequencies. Consequently, the present structure can be used as a THz grating coupler as in [30], where a wave of a certain frequency is subtracted from the substrate depending on the angle of incidence.

At 1.11 THz the field is coupled to the first mode of the substrate, while at 1.30 THz the field is coupled to multiple modes. In the latter case, the SPR is seen to be excited around the cylinder. This can lead to an extension of its effective waveguide (substrate) width. In this case, it is likely that (8.1) is satisfied for multiple diffraction orders with a wide waveguide width, exciting multiple modes as shown in Fig. 9.6d. At 1.11 and 1.30 THz, the excited SPR gives a gradual response as shown in Fig. 8.5.

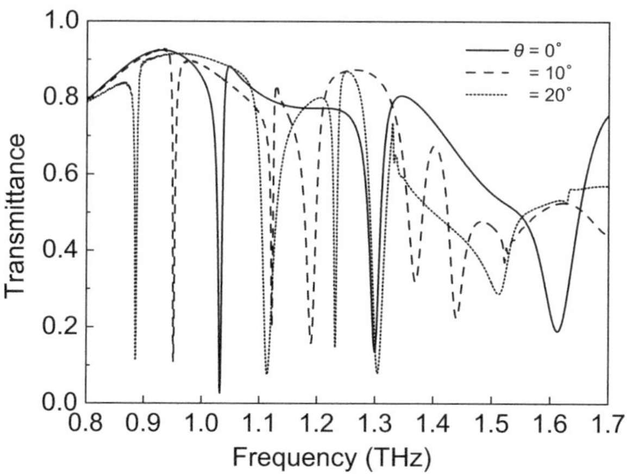

Fig. 8.5 Transmission characteristics at oblique incidence. Copyright © 2020 IEICE from [25]

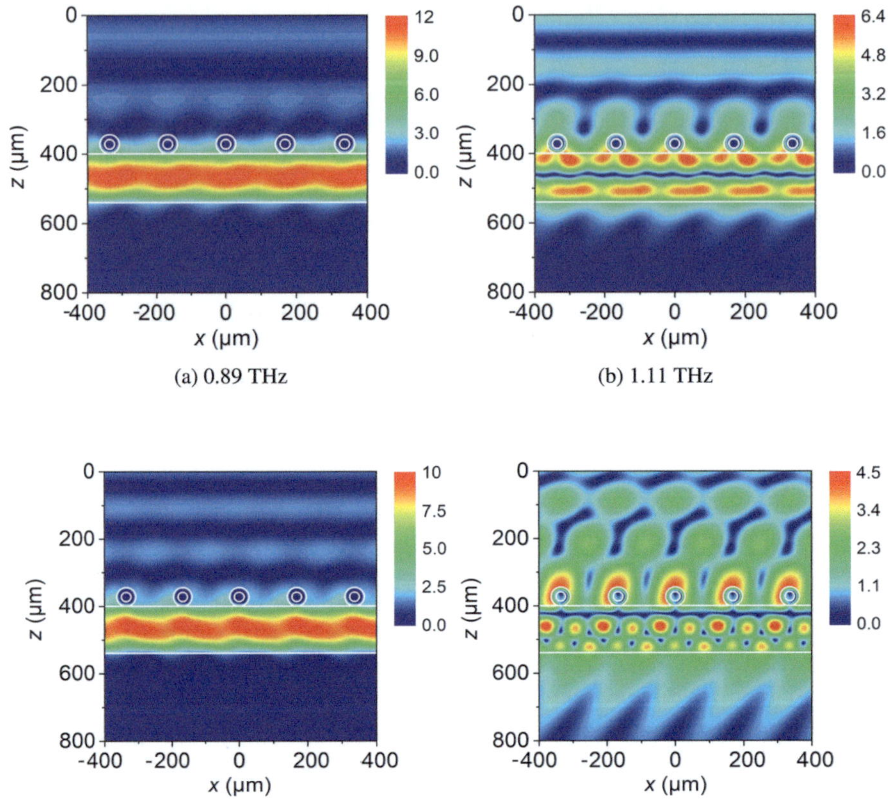

(a) 0.89 THz (b) 1.11 THz

(c) 1.23 THz (d) 1.30 THz

Fig. 8.6 Transmission characteristics at normal incidence. Copyright © 2020 IEICE from [25]

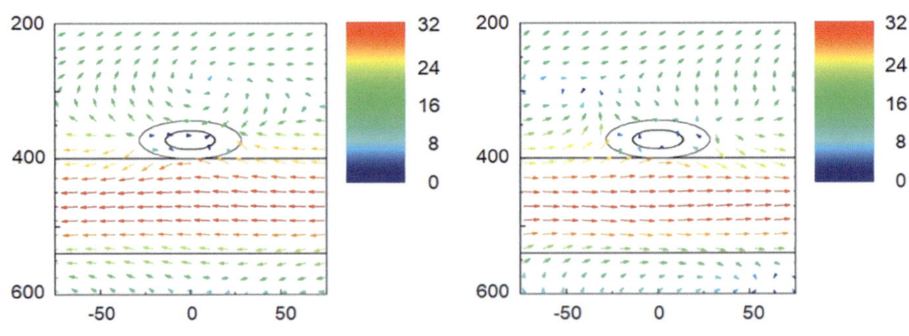

(a) 0.89 THz (b) 1.23 THz

Fig. 8.7 Poynting vector at oblique incidence

8.3 Analysis of a Finite Array of Cylinders

In the previous section, it was shown that the field propagating from the background space is coupled to the substrate. Conversely, as the field propagates through the substrate, it must radiate back to the background space. To investigate this effect, we calculate the radiation patterns for the finite array of cylinders. Figure 8.8 shows the configuration where the fundamental mode of the substrate is launched in the $+x$ direction.

The radiation patterns are shown in Fig. 8.9, where the frequency is chosen to be 0.89, 1.03, and 1.23 THz, and the number of cylinders is chosen to be $N = 10$, 20, and 50. At 0.89 THz, the field is seen to radiate in the $\theta = -20°$ directions, and the pattern becomes narrower as the number of cylinders is increased. This radiation direction is consistent with the results of Fig. 8.6a. At 1.03 THz, the field radiates toward $\theta = 0°$ in Fig. 8.9b, which agrees with the result of Fig. 8.3a. At 1.23 THz the field radiates towards $\theta = 20°$ in Fig. 8.9c, which agrees with the result of Fig. 8.6c.

Note in Fig. 8.9 that the radiation can also be seen in the $-z$ direction. This is explained by the wave vector diagram, which can be illustrated using (8.1) [31]. Figure 8.10 shows the wave vector diagram, where $2\pi/\lambda$ is the wavenumber of the background space above and below the substrate, n_s is the refractive index of the substrate, and $2\pi/\Lambda$ is the grating vector. Recall that $2\pi/\Lambda$ is constant and $2\pi/\lambda < \beta_0 < n_s 2\pi/\lambda$.

As can be seen, the direction of the beam depends on the relation between β_0 and $2\pi/\Lambda$. At 0.89 THz the beam has the wave vector component in the $-x$ direction due to $\beta_0 < 2\pi/\Lambda$. At 1.03 THz the beam has no wave component in the x direction due to $\beta_0 = 2\pi/\Lambda$. At 1.23 THz the beam has the $+x$ component because $\beta_0 > 2\pi/\Lambda$. In all cases, the diagram corresponds to a two-beam coupling [31], with two vector arrows pointing in the $\pm z$ directions. This clearly explains why the radiation occurs in the same $\pm z$ directions in Fig. 8.9. This behavior can also be seen from the field distribution for $N = 20$ shown in Fig. 8.11, where the field radiates in the $\pm z$ directions. Nevertheless, the field radiation in the direction of the substrate is hardly observed in Fig. 8.6a and c. This is probably due to the infinite periodic structure of the grating.

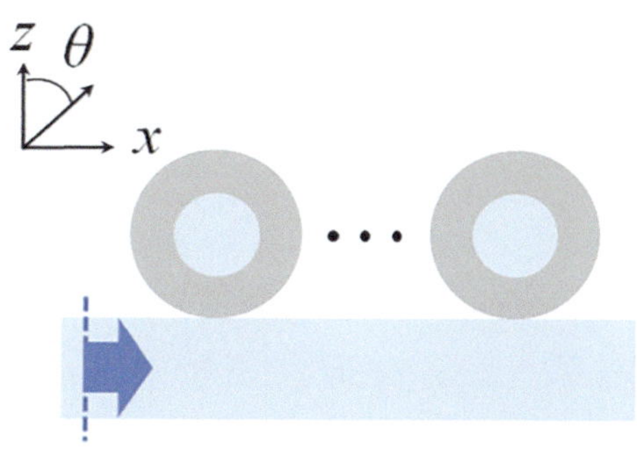

Fig. 8.8 Finite array of cylinders. Copyright © 2020 IEICE from [25]

Fig. 8.9 Radiation patterns. Copyright © 2020 IEICE from [25]

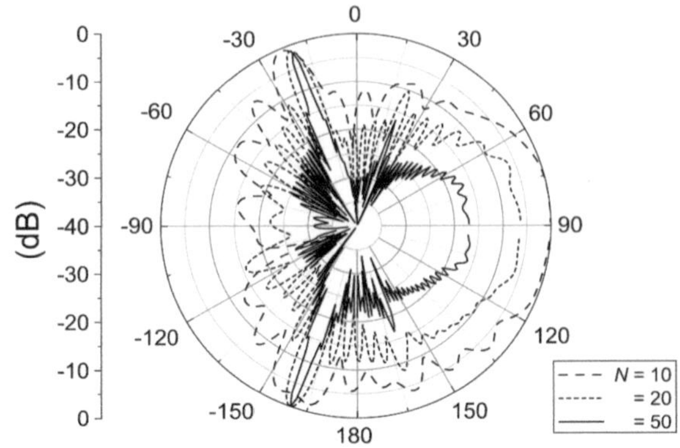

(a) 0.89 THz

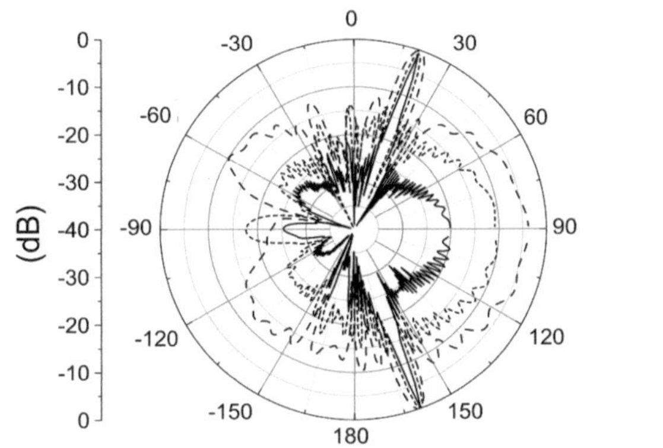

(b) 1.03 THz

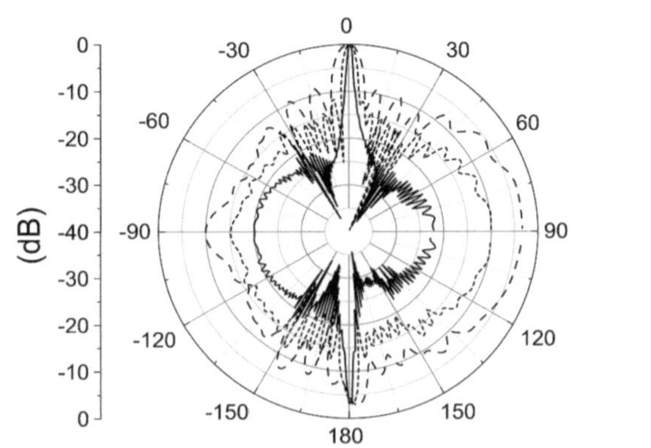

(c) 1.23 THz

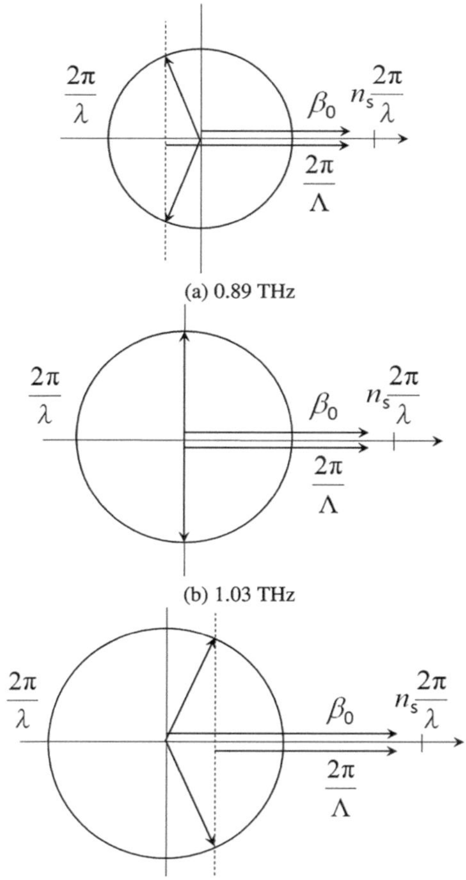

(a) 0.89 THz

(b) 1.03 THz

(c) 1.23 THz

Fig. 8.10 Wave vector diagram. Copyright © 2020 IEICE from [25]

Note that the radiation patterns are not perfectly symmetric in the $\pm z$ directions, especially at 0.89 and 1.03 THz for a small N in Fig. 8.9. This is because a small refractive index perturbation occurs on the top of the substrate due to the addition of the gratings. When another slab is added at the bottom of the structure in Fig. 8.8, only the radius of the lower half of the circle in Fig. 8.10 becomes large, changing the radiation direction in the lower half region. As a result, with a finite number of cylinders, the field propagating in the substrate can be extracted into the background space, the direction of which is controlled by the frequency of the guided mode.

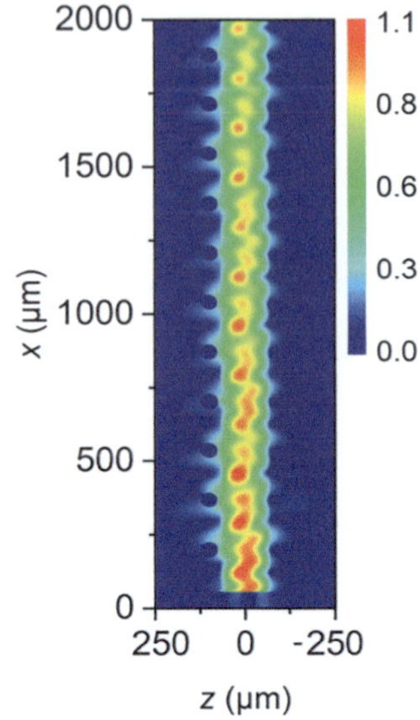

Fig. 8.11 Propagating field for the finite array at 0.89 THz

8.4 Analysis of a Infinite Array of InSb Spheres

We have analyzed 2-D gratings so far. Here, 3-D grating structures are analyzed in what follows. Figure 8.12 shows the configuration of an InSb sphere array on a substrate, where only the unit cell is shown. The period is chosen to be $\Lambda = 168\ \mu\text{m}$.

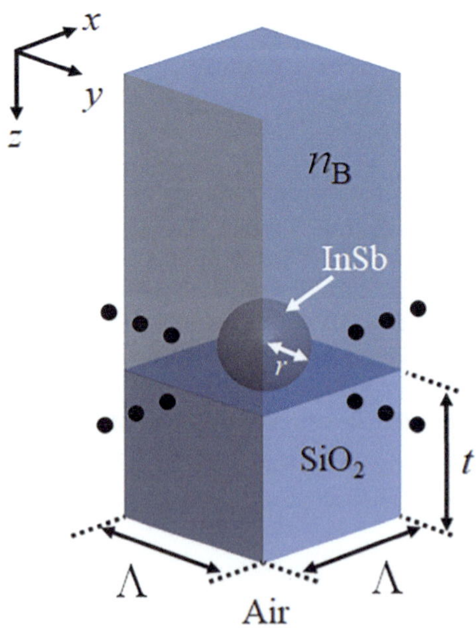

Fig. 8.12 Configuration of the unit cell of an InSb sphere array on a substrate. Copyright © 2022 IEICE from [26]

The radius of the InSb sphere is fixed to be $r = 28\ \mu\text{m}$. The material of the substrate is SiO_2 and its refractive index is set to be $n_s = 1.94$. The background refractive index above the substrate is defined by n_B. The thickness of the substrate is set to be $t = 140\ \mu\text{m}$. These structural parameters are chosen to obtain the GMR around 1 THz.

To analyze an infinite array in the x and y directions, we use the 3-D FDTD method with the periodic boundary condition. The temperature adopted here is fixed at T=300 K. The E_y (H_x) polarized plane wave at normal incidence is excited towards the $+z$ direction (with no azimuthal angle on the x-y plane). The transmission characteristics are investigated under the substrate.

Figure 8.13 shows the transmission characteristics for $n_B = 1.0$. There are three transmission dips at 0.997, 1.04 and 1.21 THz. The E_y field distributions at each frequency in the x-z and y-z planes are illustrated in Fig. 8.14a and b, respectively. As can be seen, the incident field is coupled to the field in the substrate at each frequency due to the GMR. Note that the coupled field at 0.997 THz is a fundamental mode and the fields at 1.04 and 1.21 THz are first-order modes in the substrate.

It is interesting to note that the direction of the standing wave depends on the frequency. That is, at 0.997 and 1.21 THz the standing wave is seen in the x direction, while at 1.04 THz it is seen in the y direction. To investigate the propagation direction of the coupled field in more detail, we calculate the Poynting vector at 0.997 and 1.21 THz. Figure 8.15a and b show the Poynting vector in the x-z and y-z planes at 0.977 THz, and Fig. 8.15c and d shows the vectors in the x-z and y-z planes at 1.21 TH. It is seen that the coupled field in the substrate propagates in the $\pm x$ direction in the x-z plane, while the field does not propagate laterally and oscillates vertically like the Fabry-Perot resonance in the y-z plane. Although not illustrated, the field propagates only in the y direction at 1.04 THz.

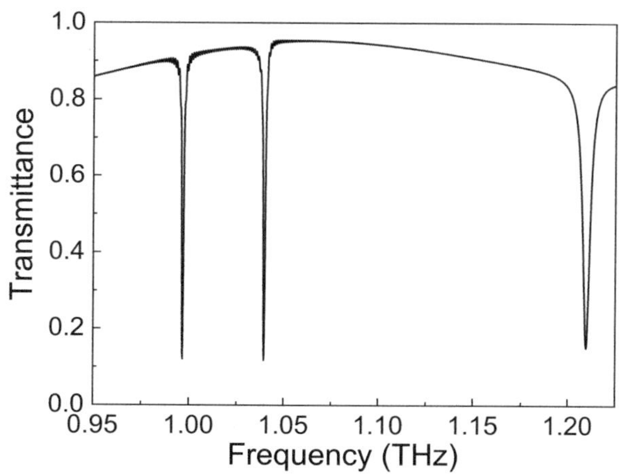

Fig. 8.13 Transmission characteristic for $n_B = 1.0$. Copyright © 2022 IEICE from [26]

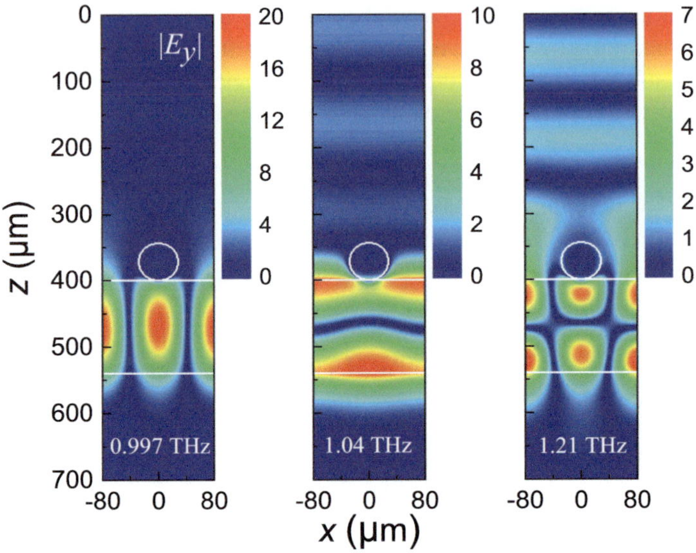

(a) *x*-*z* plane

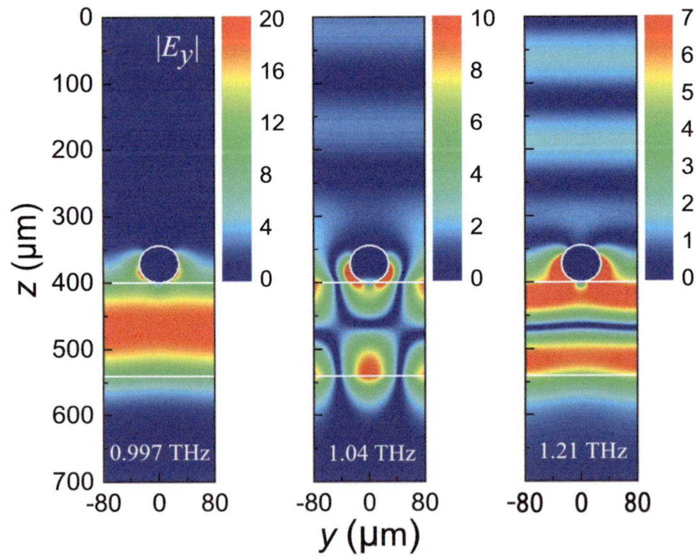

(b) *y*-*z* plane

Fig. 8.14 Field distributions. The field is normalized to the maximum amplitude of the input pulse. Copyright © 2022 IEICE from [26]

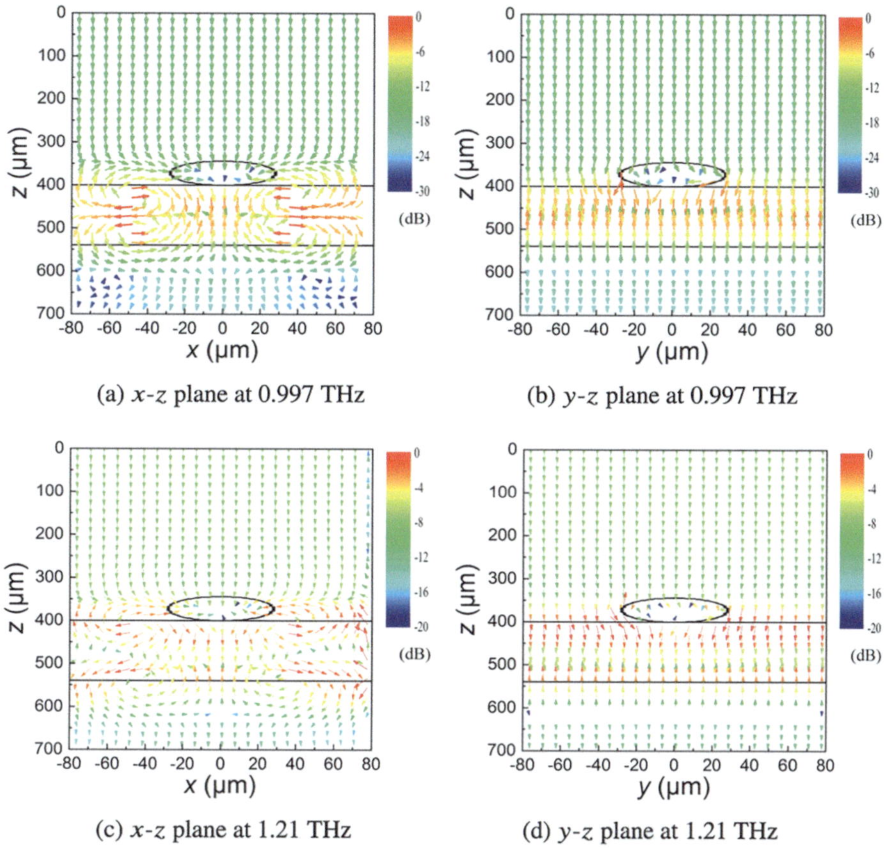

(a) x-z plane at 0.997 THz

(b) y-z plane at 0.997 THz

(c) x-z plane at 1.21 THz

(d) y-z plane at 1.21 THz

Fig. 8.15 Field distributions. The field is normalized to the maximum amplitude of the input pulse. Copyright © 2022 IEICE from [26]

The above-mentioned coupling behavior of the GMR is often explained using the following phase matching condition [32]:

$$\boldsymbol{\beta}_{\text{in}} = \boldsymbol{\beta}_0 + p\frac{2\pi}{\Lambda}\boldsymbol{x} + q\frac{2\pi}{\Lambda}\boldsymbol{y} \tag{8.2}$$

where $\boldsymbol{\beta}_{\text{in}}$ is the incident in-plane wavevector, $\boldsymbol{\beta}_0$ is the value close to the wavevector of the guided-mode in the substrate, p and q are the diffraction orders, and $\boldsymbol{x}$ and $\boldsymbol{y}$ are the unit vectors, in which $2\pi/\Lambda$ is called the grating vector. The wavevector can be decomposed into two components as $\boldsymbol{\beta}_{\text{in}} = \beta^x_{\text{in}}\boldsymbol{x} + \beta^y_{\text{in}}\boldsymbol{y}$ and $\boldsymbol{\beta}_0 = \beta^x_0\boldsymbol{x} + \beta^y_0\boldsymbol{y}$. Here, each amplitude of the vector can be determined by the azimuthal angle of the incident wave. Note, however, that the $\boldsymbol{\beta}_{\text{in}}$ of (8.2) should be ignored, since we only deal with the case at normal incidence [24, 32].

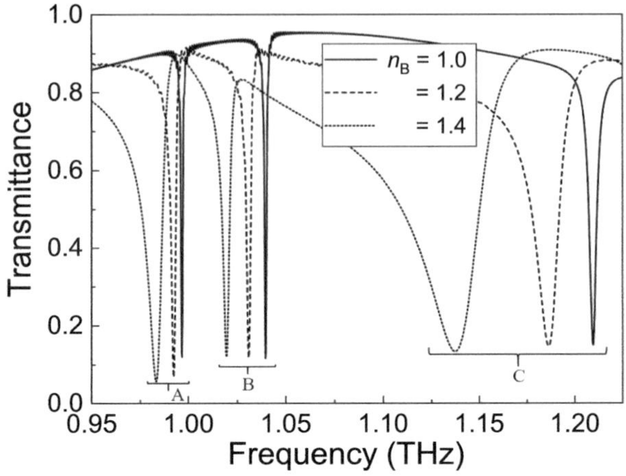

Fig. 8.16 Transmission characteristic for $n_B = 1.0$. Copyright © 2022 IEICE from [26]

As discussed in Fig. 8.15, the coupled field at 0.997 THz does not propagate laterally in the y direction, so that the grating vector in this direction (the last term in (8.2)) should be ignored, i.e., $\beta_0^x \boldsymbol{x} + p\frac{2\pi}{\Lambda}\boldsymbol{x} = 0$. It follows that the wavevector of the fundamental mode in the substrate is quite close to the grating vector in the x direction for $p = \pm 1$. Similarly, at 1.21 THz, the wavevector of the first-order mode is close to the grating vector in the x direction. The aforementioned dip frequencies resulting from the GMR can be predicted with the value of β_0^x, but this requires solving the characteristic equation of the slab waveguide.

At 1.04 THz, on the other hand, the field does not propagate laterally in the x direction (center of Fig. 8.14a). In this case, the wavevector of the first-order mode is much smaller than the grating vector in the y direction. However, the field is coupled to the substrate as shown in the center of Fig. 8.14b. This is probably because the value of the wavevector is increased by the SPP localized under the sphere shown in the center of Fig. 8.14b. Therefore, this increased wavevector in the y direction, β_0^y, is assumed to coincide with the grating vector in the same direction.

As mentioned above, the use of InSb leads to the excitation of the SPP. For the present analysis, the E_y component is the one that excites the SPP. In fact, in Fig. 8.14 (b), the SPP can be observed under the sphere at each frequency. It should be noted that the field of the SPP at 1.21 THz extends far into the background space. Therefore, the transmission dip at 1.21 THz is expected to be more strongly affected by the refractive index of the surrounding and background material.

Figure 8.16 shows the transmission characteristics for different background refractive indices n_B. As expected, the dip frequency varies much more in band C than in bands A and B. The dip frequency shift is shown in Fig. 8.17 as n_B is varied. It can be seen that the frequency shift in band C is more than three times as large as in bands A and B at $n_B = 1.4$. Therefore, for sensor applications, it is desirable to use a frequency dip where the SPP is effectively excited.

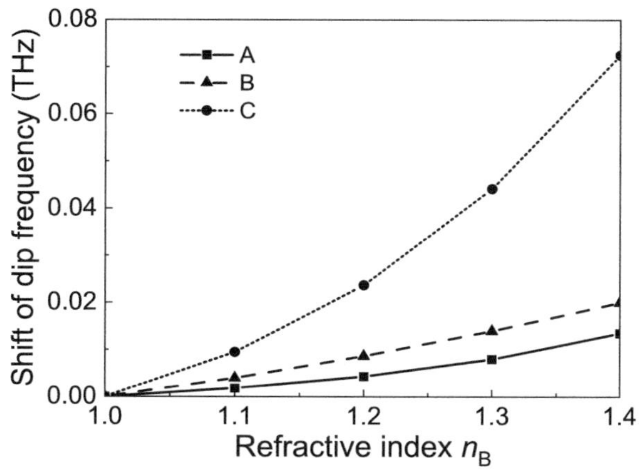

Fig. 8.17 Transmission characteristic for $n_B = 1.0$. Copyright © 2022 IEICE from [26]

8.5 Concluding Remarks

We have investigated the fundamental properties of a grating consisting of InSb-coated dielectric cylinders on a substrate using the FDTD method. The transmission characteristics for the infinite array of cylinders are calculated and discussed not only for normal incidence but also for oblique incidence. It is shown that the field propagating from the background space is coupled to the eigenmode of the substrate, leading to the use of a grating coupler at THz frequencies. A large shift of the frequency dip involving the SPR is found for the change of the background refractive index, the sensitivity of which is about ten times as large as that of the dip from the GMR alone. We have also studied a finite array of cylinders. Numerical results show that the field propagating in the substrate radiates into the background space. The direction of the radiation field can be controlled by the frequency of the guided mode. The wave vector diagram explains the relationship between the direction of the radiation field and the propagation constants.

The transmission characteristics of an InSb sphere array on a dielectric substrate have also been studied using the periodic 3-D FDTD method. It is shown that the incident field is coupled to the eigenmode in the substrate due to the GMR. In particular, the propagation direction of the coupled field is found to depend on the frequency of the transmission dip, which cannot be observed in the 2-D problem. This behavior is explained in terms of the phase matching condition. The effect of the background refractive index on the transmission characteristics is also investigated. Numerical results show that the dip frequency involving the SPP shifts significantly due to its high sensitivity to the refractive index of the surrounding and background material.

References

1. D. Taillaert, P. Bienstman, R. Baets, Compact efficient broadband grating coupler for silicon-on-insulator waveguides. Opt. Lett. **29**(23), 2749–2751 (2004)
2. K. Kintaka, Y. Kita, K. Shimizu, H. Matsuoka, S. Ura, J. Nishii, Cavity-resonator-integrated grating input/output coupler for high-efficiency vertical coupling with a small aperture. Opt. Lett. **35**(12), 1989–1991 (2010)
3. S. Ura, K. Shimizu, Y. Kita, K. Kintaka, J. Inoue, Y. Awatsuji, Integrated-optic free-space-wave coupler for package-level on-board optical interconnects. IEEE J. Sel. Top. Quantum **17**(3), 590–596 (2011)
4. J. Kang, Y. Atsumi, Y. Hayashi, J. Suzuki, Y. Kuno, T. Amemiya, N. Nishiyama, S. Arai, Amorphous-silicon inter-layer grating couplers with metal mirrors toward 3-D interconnection. IEEE J. Sel. Top. Quantum **20**(4), 8202308 (2014)
5. N. Nishiyama, J. Kang, Y. Kuno, K. Itoh, Y. Atsumi, T. Amemiya, S. Arai, Si-photonics-based layer-to-layer coupler toward 3D optical interconnection. IEICE Trans. Electron. **E101-C**(7), 501–508 (2018)
6. H.A. Haus, Y. Lai, Narrow-band optical channel-dropping filter. J. Light. Technol. **10**(1), 57–62 (1992)
7. J.M. Castro, D.F. Geraghty, S. Honkanen, C.M. Greiner, D. Iazikov, T.W. Mossberg, Demonstration of mode conversion using anti-symmetric waveguide Bragg gratings. Opt. Exp. **13**(11), 4180–4184 (2005)
8. D. Taillaert, H. Chong, P.I. Borel, L.H. Frandsen, R.M. De La Rue, R. Baets, A compact two-dimensional grating coupler used as a polarization splitter. IEEE Photonics Technol. Lett. **15**(9), 1249–1251 (2003)
9. J. Homola, S.S. Yee, G. Gauglitz, Surface plasmon resonance sensors: review. Sens. Actuators B **54**(1–2), 3–15 (1999)
10. V. Jandieri, P. Meng, K. Yasumoto, Y. Liu, Scattering of light by gratings of metal-coated circular nanocylinders on a dielectric substrate. J. Opt. Soc. Am. A **32**(7), 1384–1389 (2015)
11. K. Yasumoto, V. Jandieri, P. Meng, Y. Liu, Scattering of light by periodic array of metal-coated nanocylinders on dielectric slab, in *International Symposium on Antennas and Propagation*, vol. 3B2-1, pp. 572–573 (2016)
12. V. Jandieri, L. Okropiridze, K. Yasumoto, D. Erni, J. Pištora, Efficient analysis method of light scattering by a grating of plasmonic nanorods. COMPEL **37**(4), 1436–1448 (2018)
13. S.A. Maier, S.R. Andrews, L. Martín-Moreno, and F.J. García-Vidal, Terahertz surface plasmon-polariton propagation and focusing on periodically corrugated metal wires. Phys. Rev. Lett. **97**(17), 176805 (2006)
14. J.Y. Yin, J. Ren, H.C. Zhang, Q. Zhang, T.J. Cui, Capacitive-coupled series spoof surface plasmon polaritons. Sci. Rep. **6**, 24605 (2016)
15. J. Shibayama, J. Yamauchi, H. Nakano, Metal disc-type splitter with radially placed gratings for terahertz surface waves. Electron. Lett. **51**(4), 352–353 (2015)
16. J. Shibayama, T. Hara, M. Ito, J. Yamauchi, H. Nakano, Improved cylindrical HIE-FDTD method and its application to a metal disc-type surface wave splitter. IEEE Microw. Wirel. Compon. Lett. **29**(3), 177–179 (2019)
17. S.R.J. Bhatt, P. Bhatt, P. Deshmukh, B.R. Sangala, M.N. Satyanarayan, G. Umesh, S.S. Prabhu, Resonant terahertz InSb waveguide device for sensing polymers. J. Infrared Millim. Terahertz Waves **37**(8), 795–804 (2016)
18. X.Y. He, Q.J. Wang, S.F. Yu, Numerical study of gain-assisted terahertz hybrid plasmonic waveguide. Plasmonics **7**(3), 571–577 (2012)
19. J. Tao, B. Hu, X.Y. He, Q.J. Wang, Tunable subwavelength terahertz plasmonic stub waveguide filters. IEEE Trans. Nanotechnol. **12**(6), 1191–1197 (2013)
20. M.K. Chen, Y.C. Chang, C.E. Yang, Y. Guo, J. Mazurowski, S. Yin, P. Ruffin, C. Brantley, E. Edwards, C. Luo, Tunable terahertz plasmonic lenses based on semiconductor microslits. Microw. Opt. Technol. Lett. **52**(4), 979–981 (2010)

21. Q. Wang, Q. Tang, D. Zhang, Z. Wang, Y. Huang, Tunable terahertz spectral filter based on temperature controlled subwavelength InSb grating. Superlattices Microstruct. **75**, 955–961 (2014)
22. J. Shibayama, K. Shimizu, J. Yamauchi, H. Nakano, Surface plasmon resonance waveguide sensor in the terahertz regime. J. Light. Technol. **34**(10), 2518–2525 (2016)
23. Y. Ma, J. Zhou, J. Pištora, M. Eldlio, N. Nguyen-Huu, H. Maeda, Q. Wu, M. Cada, Subwavelength InSb-based slot wavguides for THz transport: concept and practical implementations. Sci. Rep. **6**, 38784 (2016)
24. J. Shibayama, S. Takahashi, J. Yamauchi, H. Nakano, FDTD analysis of an InSb-coated dielectric cylinder array, in *Asia-Pacific Conference on Antennas and Propagation*, vol. 6PM2-A-4, pp. 348–349 (2019)
25. J. Shibayama, S. Takahashi, J. Yamauchi, H. Nakano, Fundamental investigation of a grating consisting of InSb-coated dielectric cylinders on a substrate in the THz regime. IEICE Trans. Electron. **E103-C**(11), 567–574 (2020)
26. J. Shibayama, T. Kuroda, J. Yamauchi, H. Nakano, Analysis of an InSb sphere array on a dielectric substrate in the THz regime. IEICE Trans. Electron. **E105-C**(4), (2022)
27. A.E. Serebryannikov, K.B. Alici, E. Ozbay, A. Lakhtakia, Thermally sensitive scattering of terahertz waves by coated cylinders for tunable invisibility and masking. Opt. Exp. **26**(1), 1–14 (2018)
28. M. Tani, S. Matsuura, K. Sakai, S. Nakashima, Emission characteristics of photoconductive antennas based on low-temperature-grown GaAs and semi-insulating GaAs. Appl. Opt. **36**(30), 7853–7859 (1997)
29. M. Beck, H. Schäfer, G. Klatt, J. Demsar, S. Winnerl, M. Helm, T. Dekorsy, Impulsive terahertz radiation with high electric fields from an amplifier-driven large-area photoconductive antenna. Opt. Exp. **18**(9), 9251–9257 (2010)
30. T. Nagatsuma, S. Hisatake, M. Fujita, H.H.N. Pham, K. Tsuruda, S. Kuwano, J. Terada, Millimeter-wave and terahertz-wave applications enabled by photonics. IEEE J. Quantum Electron. **52**(1), 0600912 (2016)
31. H. Nishihara, M. Haruna, T. Suhara, *Hikari Shuseki Kairo (Optical Integrated Circuit)* (Ohm Sha, Tokyo, 1993)
32. S. Wu, J. Wang, J. Liu, H. Zhu, P. Zhang, Optical polarization conversion based on guided-mode resonance through a silicon film perforated by L-shaped holes. Opt. Mater. Express **8**(8), 2395–2402 (2018)

Chapter 9
Terahertz Surface Wave Splitter

Abstract This capter presents a novel method for splitting terahertz (THz) surface waves based on the propagation characteristics of spoof surface plasmon plaritons. Metal disk splitters with a central hole and radially arranged gratings are proposed and analyzed using the FDTD method in cylindrical coordinates. Numerical results show that the disks act as bi- and multidirectional THz surface wave splitters with large coupling regions, the frequency of which is selected by the grating structures. To efficiently analyze cylindrical structures, we further develop the cylindrical hybrid implicit-explicit finite-difference time-domain method, in which the implicit scheme is adopted only in the circumferential direction. The connection of the field in a specific plane gives the cyclic matrix problem, the solution of which is performed with the Sherman-Morrison formula. The convolutional perfectly matched layer is also introduced in cylindrical coordinates. The present method is applied to the analysis of the metal disk type surface wave splitter. The computational time is reduced to less than 20% of the explicit method, yielding almost the same numerical results. The calculated radiation patterns are compared with experimental results at gigahertz frequencies.

9.1 Introduction

A surface plasmon polariton (SPP) is known to propagate along a metal-insulator interface at optical frequencies, but not at terahertz (THz) frequencies. Note that a spoof SPP can be guided along metal grating structures at THz frequencies [1]. Using the characteristics of the spoof SPP, a bi-directional splitter has been numerically studied using the two-dimensional (2-D) finite-difference time-domain (FDTD) method [2]. Bi- and multidirectional splitters consisting of finite-thickness metal gratings have also been studied numerically and experimentally [3, 4]. Unfortunately, the wave coupling in these splitters is limited to the narrow regions of the grating. This fact inspires us to consider a splitter structure with larger coupling regions.

Here, we propose simple THz frequency splitters consisting of metal disks with a center hole and radially arranged gratings [5]. In the proposed splitters, the gratings are located in fan-shaped regions to which THz waves are widely coupled. In addition,

J. Shibayama, *FDTD Analysis of Plasmonic Devices*, Springer Series in Optical Sciences 254, https://doi.org/10.1007/978-981-96-8130-3_9

since the grooves of the gratings are terminated in the circumferential direction, the waves are effectively confined to the grating regions. The splitters are analyzed using the FDTD method in cylindrical coordinates. Numerical results show that the disks act as bi- and multidirectional THz wave splitters with large coupling regions.

Incidentally, a hybrid implicit-explicit (HIE) FDTD method has received much attention as an efficient method for the problem where only one specific spatial direction needs to be discretized with a small mesh [6–11]. For the analysis of cylindrical structures, the HIE-FDTD method has been formulated in cylindrical coordinates [12]. Since the circumferential mesh size inevitably becomes small near the center axis, the implicit scheme is successfully introduced only in the circumferential direction.

Note that in cylindrical coordinates, the field is considered to be periodic in the circumferential direction. It follows that the field should be connected in a certain plane in this direction. Unfortunately, the application of the implicit scheme gives rise to the cyclic matrix problem, the solution of which was not discussed in [12]. In addition, the application is limited to a simple cylindrical cavity resonator, which allows the calculation without an absorbing boundary condition.

We make the cylindrical HIE-FDTD method [12] a usable technique for treating general electromagnetic problems in cylindrical structures. To efficiently solve the cyclic matrix, we take advantage of the Sherman-Morrison formula [13, 14] which allows the use of the Thomas algorithm. In addition, the convolutional perfectly matched layer (CPML) [15] is adopted in cylindrical coordinates [16]. As an application, we analyze a metal disk type surface wave splitter at GHz frequencies [17]. Here, we reexamine a splitter at GHz frequencies with an incident linearly polarized wave for the fabrication of the device. It is found that the field distributions on the disk surface and the radiation patterns are in good agreement with those obtained by the explicit FDTD method. The computational time is reduced to less than 20% compared to the explicit method, while maintaining almost the same accuracy. The calculated radiation patterns are also compared with experimental results. A good agreement between them is obtained, demonstrating the validity of the analysis.

9.2 Design and Analysis of the Splitter

For the design of a splitter, we use the dispersion relations (TM-polarized waves) for one-dimensional groove arrays as [18, 19]

$$k = \frac{\omega}{c}\sqrt{\frac{d^2}{p^2}\tan^2\left(\frac{\omega}{c}h\right) + 1} \tag{9.1}$$

where k is the wavenumber, p is the grating period, d is the groove width, h is the groove depth, c is the speed of light in a vacuum and ω is the angular frequency. Figure 9.1 shows the dispersion curves for $d = 20\ \mu\text{m}$ and $p = 50\ \mu\text{m}$ with different

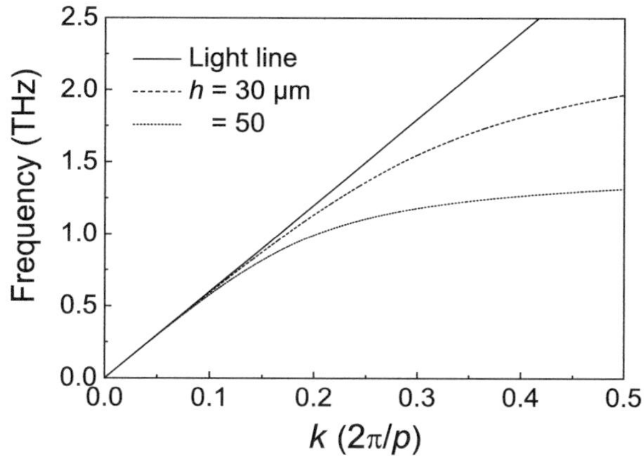

Fig. 9.1 Dispersion curves. From [5] Copyright © 2015 by John Wiley Sons, Inc. Reproduced by permission of John Wiley & Sons, Inc

groove depths for $h = 30$ and $50\ \mu\text{m}$. It is expected that the electromagnetic surface wave at 1.5 THz is coupled to the gratings for $h = 30\ \mu\text{m}$, while that at 1 THz is coupled to those for $h = 50\ \mu\text{m}$. Note that part of the wave at 1 THz can also be coupled to the gratings for $h = 30\ \mu\text{m}$, since the dispersion relation is satisfied. We adopt these two groove depths for the design of a bidirectional splitter.

Figure 9.2 shows a bidirectional splitter consisting of a metal disk with a central hole and radially arranged gratings, where a radially polarized beam is launched from the bottom of the hole. The groove depths are chosen to be $h = 30$ and $50\ \mu\text{m}$ on the left and right sides, respectively, where $d = 20\ \mu\text{m}$, $p = 50\ \mu\text{m}$, $r = 130\ \mu\text{m}$, $\varphi_L =$

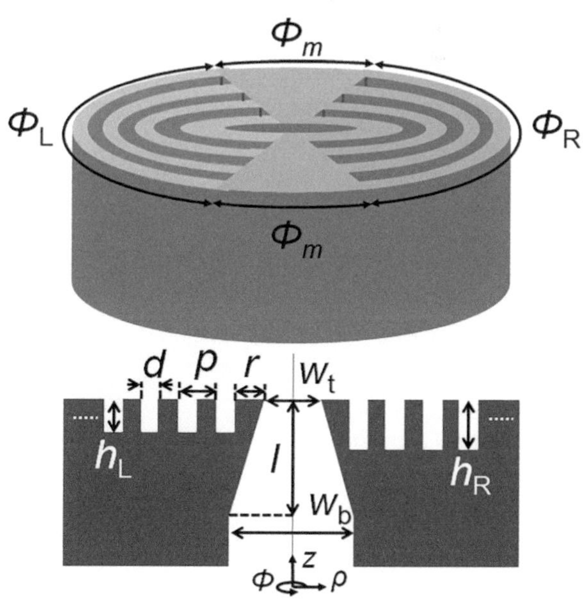

Fig. 9.2 Configuration of bidirectional splitter. From [5] Copyright © 2015 by John Wiley Sons, Inc. Reproduced by permission of John Wiley & Sons, Inc

$\varphi_R = 135°$, and $\varphi_m = 45°$. The radii of the center hole are $w_t = 200$ μm and $w_b = 480$ μm, and the disk thickness is $l = 400$ μm. The FDTD method in cylindrical coordinates is used, with $\Delta z = \Delta\rho = 2$ μm and $\Delta\varphi = 15°$. Outside the disk, the CPML is imposed to absorb outgoing waves.

The treatment of metal at THz frequencies in practical simulations [20] must be done carefully, i.e. either a perfect electric conductor (PEC) or a dispersive material is assumed. The choice of one depends on the structure and the frequency to be studied [20]. Therefore, we have preliminarily analyzed the splitter using the PEC model and the Drude model for Ag, the latter of which takes into account the metal dispersion using the frequency-dependent formulation. It is shown that the two results are almost identical. A similar observation can be found in [10], where spoof SPPs in 2-D grating structures are analyzed using the FDTD method. This is because the resonance in the gratings is relatively weak and the THz frequencies are not so high (less than 2 THz) [20]. Therefore, we use the PEC model in the following analysis.

The H_ϕ components at the top of the disk are shown in Fig. 9.3. The bidirectional nature of the beam splitting is clearly seen at 1 and 1.5 THz. It is noteworthy that the THz waves extend far into the fan-shaped grating regions. This splitter can also be used for multi-channel biosensing components [22], which have the large interaction region of an analyte and THz waves. From Fig. 9.3a, it can be seen that most of the surface wave at 1 THz propagates to the right, while some propagates to the left. This is due to the fact that the surface wave at 1 THz is also coupled to the left gratings for $h = 30$ μm, as expected from the dispersion relation in Fig. 9.1. On the other hand, the field decrease at 1.5 THz is pronounced in the right gratings, as can be seen in Fig. 9.3b. This is because the surface wave at 1.5 THz is in the cutoff condition for $h = 50$ μm, as found in Fig. 9.1.

The power propagating in each direction is calculated. The radiated power to the top of the disk (P_{rad}) and the reflected power to the bottom (P_{ref}) are calculated as

$$P_{\text{rad,ref}} = \frac{1}{2}\iint \text{Re}\left(E_\rho H_\phi^* - E_\phi H_\rho^*\right) ds \tag{9.2}$$

In this calculation, the integral calculation of P_{rad} is performed at the ρ-ϕ plane 100 μm above the metal surface, while that of P_{ref} is performed at the ρ-ϕ plane (radius of $\rho = w_b/2$) 75 μm below the incidence plane.

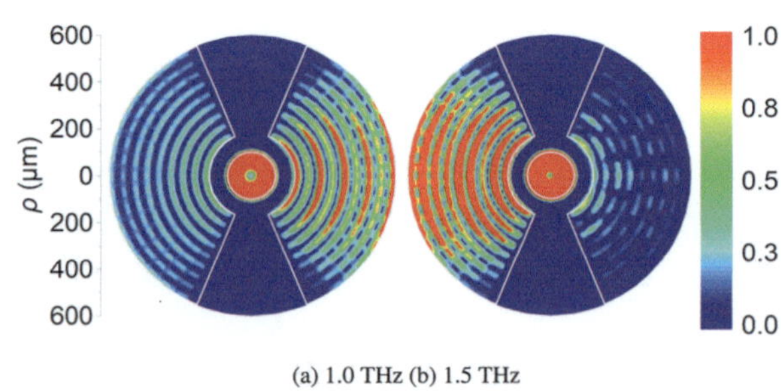

Fig. 9.3 H_ϕ components on the metal surface. From [5] Copyright © 2015 by John Wiley Sons, Inc. Reproduced by permission of John Wiley & Sons, Inc

The power propagating along the gratings in the right and left directions ($P_{\mathrm{right,left}}$) is calculated as

$$P_{\mathrm{left,right}} = \frac{1}{2}\iint \mathrm{Re}\left(E_z H_\phi^* - E_\phi H_z^*\right) ds \tag{9.3}$$

In this calculation, the integral calculation is performed in the z-ϕ plane with a radius of $\rho = 600\ \mu\mathrm{m}$, where the integration area in the z direction is from the metal surface to $z = 100\ \mu\mathrm{m}$, where P_{rad} is calculated. $P_{\mathrm{right,left}}$ are the integrals of the z-ϕ plane 180° to the right and 180° to the left, respectively.

The power propagating in each direction is evaluated using (9.2) and (9.3). First, the power at 1.0 THz is considered. As a result, 7% of the power is radiated to the top of the disk and 85% is reflected, with 5% of the power traveling to the right and 3% to the left. Figure 9.4a shows the H_ϕ component of the ρ-z region. Since the radially polarized eigenmode is excited in the +z direction just before the start of

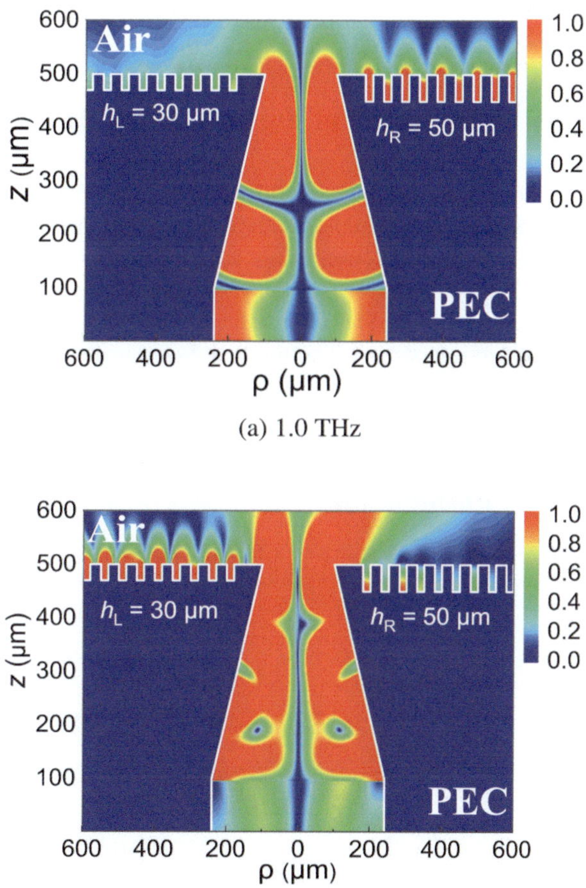

(a) 1.0 THz

(b) 1.5 THz

Fig. 9.4 Transmission characteristics at normal incidence the tapered structure in the input section, only the reflection field component can be observed in the $z < 100$ μm region. As can be seen from the power evaluation, a lot of power is reflected. This is because the 1.0 THz radial polarization mode is a cutoff condition at $w_t = 200$ μm. The reflections at the top of the taper due to the cutoff create standing waves at the input. As mentioned above, surface waves are also coupled to the grating on the left side.

Next, we discuss the power at 1.5 THz. In this case, 68% of the power is radiated and 23% is reflected, with 8% of the power going to the left side and 1% to the right side. 1.5 THz does not satisfy the dispersion relation for the right grating, so the fields do not combine, and Fig. 9.4b shows that many field components radiate upward.

It should be mentioned that the power coupled to the gratings is relatively low, i.e. a few percent of the incident power as discussed above. This is because most of the power propagates through the hole to the upper air region, as in conventional splitters. In order to increase the coupled power, a cover plate with or without gratings can be placed over the current splitters. In addition, the use of the cylindrical wire to excite surface waves [3] may be effective even in the present splitters.

9.3 Analysis Using the Cylindrical HIE-FDTD Method

Note that the sampling width of the circumferential direction in cylindrical coordinates inevitably becomes small near the center axis. This leads to a small time step due to the CFL condition, which results in a long computation time. To alleviate this problem, we introduce the implicit technique only in the circumferential direction in cylindrical coordinates, resulting in a cylindrical HIE-FDTD method. In this section, we analyze the disk type surface wave splitter using the cylindrical HIE-FDTD method [23].

9.3.1 Formulation

Maxwell's equations in cylindrical coordinates are expressed as

$$-\mu_r \frac{1}{c}\frac{\partial H_\rho}{\partial t} = \frac{1}{\rho}\frac{\partial E_z}{\partial \phi} - \frac{\partial E_\phi}{\partial z} \tag{9.4a}$$

$$-\mu_r \frac{1}{c}\frac{\partial H_\phi}{\partial t} = \frac{\partial E_\rho}{\partial z} - \frac{\partial E_z}{\partial \rho} \tag{9.4b}$$

$$-\mu_r \frac{1}{c}\frac{\partial H_z}{\partial t} = \frac{1}{\rho}\frac{\partial (\rho E_\phi)}{\partial \rho} - \frac{1}{\rho}\frac{\partial E_\rho}{\partial \phi} \tag{9.4c}$$

$$\varepsilon_r \frac{1}{c}\frac{\partial E_\rho}{\partial t} = \frac{1}{\rho}\frac{\partial H_z}{\partial \phi} - \frac{\partial H_\phi}{\partial z} \tag{9.4d}$$

$$\varepsilon_r \frac{1}{c}\frac{\partial E_\phi}{\partial t} = \frac{\partial H_\rho}{\partial z} - \frac{\partial H_z}{\partial \rho} \tag{9.4e}$$

$$\varepsilon_r \frac{1}{c}\frac{\partial E_z}{\partial t} = \frac{1}{\rho}\frac{\partial (\rho H_\phi)}{\partial \rho} - \frac{1}{\rho}\frac{\partial H_\rho}{\partial \phi} \tag{9.4f}$$

where c is the speed of light in a vacuum, ε_r is the relative permittivity, and μ_r is the relative permeability. Note that E and H are normalized to equalize their amplitudes. We first show the fundamental equations as

$$H_\rho^{n+1/2} = H_\rho^{n-1/2} - \frac{c\Delta t}{\mu_r}\left[\frac{1}{\rho}\frac{\partial E_z^n}{\partial \phi} - \frac{\partial E_\phi^n}{\partial z}\right] \tag{9.5a}$$

$$H_\phi^{n+1/2} = H_\phi^{n-1/2} - \frac{c\Delta t}{\mu_r}\left[\frac{\partial E_\rho^n}{\partial z} - \frac{\partial E_z^n}{\partial \rho}\right] \tag{9.5b}$$

$$H_z^{n+1/2} = H_z^{n-1/2} - \frac{c\Delta t}{\mu_r}\left[\frac{1}{\rho}\frac{\partial (\rho E_\phi^n)}{\partial \rho} - \frac{1}{\rho}\frac{\partial E_\rho^n}{\partial \phi}\right] \tag{9.5c}$$

$$E_\rho^{n+1} = E_\rho^n + \frac{c\Delta t}{\varepsilon_r}\left[\frac{1}{\rho}\frac{\partial H_z^{n+1/2}}{\partial \phi} - \frac{\partial H_\phi^{n+1/2}}{\partial z}\right] \tag{9.5d}$$

$$E_\phi^{n+1} = E_\phi^n + \frac{c\Delta t}{\varepsilon_r}\left[\frac{\partial H_\rho^{n+1/2}}{\partial z} - \frac{\partial H_z^{n+1/2}}{\partial \rho}\right] \tag{9.5e}$$

$$E_z^{n+1} = E_z^n + \frac{c\Delta t}{\varepsilon_r}\left[\frac{1}{\rho}\frac{\partial (\rho H_\phi^{n+1/2})}{\partial \rho} - \frac{1}{\rho}\frac{\partial H_\rho^{n+1/2}}{\partial \phi}\right] \tag{9.5f}$$

Here, we introduce the implicit technique only in the circumferential direction, where the time average is applied to the ϕ derivatives as

$$H_\rho^{n+1} = H_\rho^n - \frac{c\Delta t}{\mu_r}\left[\frac{1}{\rho}\frac{\partial}{\partial \phi}\left(\frac{E_z^{n+1} + E_z^n}{2}\right) - \frac{\partial E_\phi^{n+1/2}}{\partial z}\right] \tag{9.6a}$$

$$H_\phi^{n+1/2} = H_\phi^{n-1/2} - \frac{c\Delta t}{\mu_r}\left[\frac{\partial E_\rho^n}{\partial z} - \frac{\partial E_z^n}{\partial \rho}\right] \tag{9.6b}$$

$$H_z^{n+1} = H_z^n - \frac{c\Delta t}{\mu_r}\left[\frac{1}{\rho}\frac{\partial (\rho E_\phi^{n+1/2})}{\partial \rho} - \frac{1}{\rho}\frac{\partial}{\partial \phi}\left(\frac{E_\rho^{n+1} + E_\rho^n}{2}\right)\right] \tag{9.6c}$$

$$E_\rho^{n+1} = E_\rho^n + \frac{c\Delta t}{\varepsilon_r}\left[\frac{1}{\rho}\frac{\partial}{\partial \phi}\left(\frac{H_z^{n+1} + H_z^n}{2}\right) - \frac{\partial H_\phi^{n+1/2}}{\partial z}\right] \tag{9.6d}$$

$$E_\phi^{n+1/2} = E_\phi^{n-1/2} + \frac{c\Delta t}{\varepsilon_r}\left[\frac{\partial H_\rho^n}{\partial z} - \frac{\partial H_z^n}{\partial \rho}\right] \tag{9.6e}$$

$$E_z^{n+1} = E_z^n + \frac{c\Delta t}{\varepsilon_r}\left[\frac{1}{\rho}\frac{\partial (\rho H_\phi^{n+1/2})}{\partial \rho} - \frac{1}{\rho}\frac{\partial}{\partial \phi}\left(\frac{H_\rho^{n+1} + H_\rho^n}{2}\right)\right] \tag{9.6f}$$

Focusing on (9.6a), (9.6c), (9.6d), and (9.6f), we notice that the time indexes of H_ρ, H_z, E_ρ, and E_z are at n and $n+1$. In addition, those of H_ϕ and E_ϕ are at $n-1/2$ and $n+1/2$. First, we explicitly solve (9.6b) and (9.6e). Next, we substitute (9.6c) into (9.6d) and (9.6a) into (9.6f), and then solve them implicitly. Finally, solve (9.6a) and (9.6c) explicitly.

We will discuss the treatment of the field on the central axis in cylindrical coordinates. Note that the E_z field component can be set to zero if the structure is circularly symmetric and the field oscillates sinusoidally in the circumferential direction. However, if the structure is not circularly symmetric, we must calculate the E_z component by solving the following Ampere law:

$$\frac{\varepsilon_r}{c}\int_S \frac{\partial \boldsymbol{E}}{\partial t} = \oint_C \boldsymbol{H}\cdot \rho d\phi \tag{9.7}$$

where C is a closed circle including the z axis and S is the boundary of the circle. For C to be $\frac{\Delta\rho}{2}$, we get

$$E^{n+1}_{z(0,j,k+1/2)} = E^{n}_{z(0,j,k+1/2)} + \frac{2c\Delta t\Delta\phi}{\varepsilon_r\pi\Delta\rho}\sum_{p=0}^{N_\phi-1} H^{n+1/2}_{\phi(1/2,p,k+1/2)} \tag{9.8}$$

where, N_ϕ is the number of sampling points in the circumferential direction. Equation (9.8) is solved explicitly.

The application of the boundary condition is also important for open region problems. The CPML with the complex frequency shifted parameters is applied to the cylindrical HIE-FDTD method. The resultant equations are as follows:

$$\begin{aligned}
H^{n+1}_{\rho(i,j+1/2,k+1/2)} &= H^{n}_{\rho(i,j+1/2,k+1/2)} \\
&\quad - \frac{c\Delta t}{2\mu_r\rho_i\Delta\phi}\Big[E^{n+1}_{z(i,j+1,k+1/2)} - E^{n+1}_{z(i,j,k+1/2)} \\
&\qquad + E^{n}_{z(i,j+1,k+1/2)} - E^{n}_{z(i,j,k+1/2)}\Big] \\
&\quad + \frac{c\Delta t}{\mu_r\kappa_z\Delta z}\Big[E^{n+1/2}_{\phi(i,j+1/2,k+1)} - E^{n+1/2}_{\phi(i,j+1/2,k)}\Big] \\
&\quad + \frac{c\Delta t}{\mu_r}\Psi^{n+1/2}_{h\rho z(i,j+1/2,k+1/2)}
\end{aligned} \tag{9.9a}$$

$$\begin{aligned}
H^{n+1/2}_{\phi(i+1/2,j,k+1/2)} &= H^{n-1/2}_{\phi(i+1/2,j,k+1/2)} \\
&\quad - \frac{c\Delta t}{\mu_r\kappa_z\Delta z}\Big[E^{n}_{\rho(i+1/2,j,k+1)} - E^{n}_{\rho(i+1/2,j,k)}\Big] \\
&\quad + \frac{c\Delta t}{\mu_r\kappa_\rho\Delta\rho}\Big[E^{n}_{z(i+1,j,k+1/2)} - E^{n}_{z(i,j,k+1/2)}\Big] \\
&\quad - \frac{c\Delta t}{\mu_r}\Psi^{n}_{h\phi z(i+1/2,j,k+1/2)} + \frac{c\Delta t}{\mu_r}\Psi^{n}_{h\phi\rho(i+1/2,j,k+1/2)}
\end{aligned} \tag{9.9b}$$

$$\begin{aligned} H^{n+1}_{z(i+1/2,j+1/2,k)} &= H^{n}_{z(i+1/2,j+1/2,k)} \\ &- \frac{c\Delta t}{\mu_r \kappa_\rho \rho_{i+1/2}\Delta\rho}\left[\rho_{i+1} E^{n+1/2}_{\phi(i+1,j+1/2,k)} - \rho_i E^{n+1/2}_{\phi(i,j+1/2,k)}\right] \\ &+ \frac{c\Delta t}{2\mu_r \rho_{i+1/2}\Delta\phi}\left[E^{n+1}_{\rho(i+1/2,j+1,k)} - E^{n+1}_{\rho(i+1/2,j,k)} \right.\\ &\qquad\left. + E^{n}_{\rho(i+1/2,j+1,k)} - E^{n}_{\rho(i+1/2,j,k)}\right] \\ &- \frac{c\Delta t}{\mu_r}\Psi^{n+1/2}_{hz\rho(i+1/2,j+1/2,k)} \end{aligned} \tag{9.9c}$$

$$\begin{aligned} E^{n+1}_{\rho(i+1/2,j,k)} &= E^{n}_{\rho(i+1/2,j,k)} \\ &+ \frac{c\Delta t}{2\varepsilon_r \rho_{i+1/2}\Delta\phi}\left[H^{n+1}_{z(i+1/2,j+1/2,k)} - H^{n+1}_{z(i+1/2,j-1/2,k)} \right.\\ &\qquad\left. + H^{n}_{z(i+1/2,j+1/2,k)} - H^{n}_{z(i+1/2,j-1/2,k)}\right] \\ &- \frac{c\Delta t}{\varepsilon_r \kappa_z \Delta z}\left[H^{n+1/2}_{\phi(i+1/2,j,k+1/2)} - H^{n+1/2}_{\phi(i+1/2,j,k-1/2)}\right] \\ &- \frac{c\Delta t}{\varepsilon_r}\Psi^{n+1/2}_{e\rho z(i+1/2,j,k)} \end{aligned} \tag{9.9d}$$

$$\begin{aligned} E^{n+1/2}_{\phi(i,j+1/2,k)} &= E^{n-1/2}_{\phi(i,j+1/2,k)} \\ &+ \frac{c\Delta t}{\varepsilon_r \kappa_z \Delta z}\left[H^{n}_{\rho(i,j+1/2,k+1/2)} - H^{n}_{\rho(i,j+1/2,k-1/2)}\right] \\ &- \frac{c\Delta t}{\varepsilon_r \kappa_\rho \Delta\rho}\left[H^{n}_{z(i+1/2,j+1/2,k)} - H^{n}_{z(i-1/2,j+1/2,k)}\right] \\ &+ \frac{c\Delta t}{\varepsilon_r}\Psi^{n}_{e\phi z(i,j+1/2,k)} - \frac{c\Delta t}{\varepsilon_r}\Psi^{n}_{e\phi\rho(i,j+1/2,k)} \end{aligned} \tag{9.9e}$$

$$\begin{aligned} E^{n+1}_{z(i,j,k+1/2)} &= E^{n}_{z(i,j,k+1/2)} \\ &+ \frac{c\Delta t}{\varepsilon_r \kappa_\rho \rho_i \Delta\rho}\left[\rho_{i+1/2} H^{n+1/2}_{\phi(i+1/2,j,k+1/2)} - \rho_{i-1/2} H^{n+1/2}_{\phi(i-1/2,j,k+1/2)}\right] \\ &- \frac{c\Delta t}{2\varepsilon_r \rho_i \Delta\phi}\left[H^{n+1}_{\rho(i,j+1/2,k+1/2)} - H^{n+1}_{\rho(i,j-1/2,k+1/2)} \right.\\ &\qquad\left. + H^{n}_{\rho(i,j+1/2,k+1/2)} - H^{n}_{\rho(i,j-1/2,k+1/2)}\right] \\ &+ \frac{c\Delta t}{\varepsilon_r}\Psi^{n+1/2}_{ez\rho(i,j,k+1/2)} \end{aligned} \tag{9.9f}$$

where κ_ξ is the CFS parameter, and ξ is ρ or z.

9.3.2 Treatment of the Cyclic Matrix

Note that the above equation gives a cyclic matrix that should be solved implicitly. However, the cyclic matrix cannot be solved directly by the Thomas algorithm. Here we explain how to handle the cyclic matrix.

In cylindrical coordinates, the fields are considered to be periodic in the circumferential (φ) direction. Therefore, it is necessary to connect the fields at a particular plane in the φ direction using the periodic boundary condition. For the explicit FDTD method, the application of the periodic boundary condition is straightforward. For the implicit FDTD method, however, its application results in the cyclic matrix problem. This is true for the E_ρ and E_z components, the equations of which include the derivatives in the φ direction. For example, the equation to be solved for the E_ρ component is as follows:

$$[M]\boldsymbol{x} = \boldsymbol{b} \tag{9.10}$$

where $\boldsymbol{x}$ is a column vector including the unknown e_ρ component, $\boldsymbol{b}$ is that including the known E_ρ and H_z components and

$$[M] = \begin{bmatrix} -\alpha & 1 & 0 & \cdots & \cdots & \cdots & 1 \\ 1 & -\alpha & 1 & 0 & \cdots & \cdots & 0 \\ 0 & 1 & -\alpha & 1 & 0 & \cdots & 0 \\ \vdots & \ddots & \ddots & \ddots & \ddots & \cdots & \vdots \\ 0 & \cdots & 0 & 1 & -\alpha & 1 & 0 \\ 0 & \cdots & \cdots & 0 & 1 & -\alpha & 1 \\ 1 & \cdots & \cdots & \cdots & 0 & 1 & -\alpha \end{bmatrix}$$

in which

$$\alpha = 2 + 4\varepsilon_r \mu_r \rho_{i+\frac{1}{2}}^2 \Delta\varphi^2/(c\Delta t)^2.$$

Note that $[M]$ is a cyclic matrix with non-zero components in the upper right and lower left elements. Therefore, (9.10) cannot be solved directly using the Thomas algorithm for a tridiagonal system of linear equations. Then, we rewrite (9.10) as

$$[M] = [N] + \boldsymbol{w}_1 \boldsymbol{w}_2^T$$

in which

$$\boldsymbol{w}_1 = \boldsymbol{w}_2 = [1, 0, \ldots, 0, 1]^T$$

$$[N] = \begin{bmatrix} -\alpha - 1 & 1 & 0 & \cdots & \cdots & \cdots & 0 \\ 1 & -\alpha & 1 & 0 & \cdots & \cdots & 0 \\ 0 & 1 & -\alpha & 1 & 0 & \cdots & 0 \\ \vdots & \ddots & \ddots & \ddots & \ddots & \cdots & \vdots \\ 0 & \cdots & 0 & 1 & -\alpha & 1 & 0 \\ 0 & \cdots & \cdots & 0 & 1 & -\alpha & 1 \\ 0 & \cdots & \cdots & \cdots & 0 & 1 & -\alpha - 1 \end{bmatrix}.$$

Here, we utilize the Sherman-Morrison formula [13, 14], leading to

$$[M]^{-1} = [N]^{-1} - \frac{[N]^{-1}\boldsymbol{w}_1\boldsymbol{w}_2^T[N]^{-1}}{1 + \boldsymbol{w}_2^T[N]^{-1}\boldsymbol{w}_1}.$$

To solve $\boldsymbol{x}$, the following two auxiliary equations are introduced:

$$[N]\boldsymbol{x}_1 = \boldsymbol{b} \tag{9.11}$$

$$[N]\boldsymbol{x}_2 = \boldsymbol{w}_1. \tag{9.12}$$

Once $\boldsymbol{x}_1$ and $\boldsymbol{x}_2$ are available, we obtain $\boldsymbol{x}$ in (9.10) as

$$\boldsymbol{x} = \boldsymbol{x}_1 + A\boldsymbol{x}_2$$

where

$$A = -\frac{\boldsymbol{w}_2^T\boldsymbol{x}_1}{1 + \boldsymbol{w}_2^T\boldsymbol{x}_2}.$$

Fortunately, (9.11) and (9.12) are tridiagonal systems of linear equations, to which the Thomas algorithm is applicable. Since $\boldsymbol{w}_1$ is constant, $\boldsymbol{x}_2$ of (9.12) is solved once and stored throughout the analysis.

9.4 Design, Analysis and Fabrication

At terahertz frequencies, the surface wave splitter is several hundred meters in size. Here, we design the splitter at gigahertz frequencies in order to fabricate it at a practical size.

Figure 9.5 shows the dispersion relation as a function of wavenumber k, using the grating period $p = 5$ mm, the groove width $d = 2$ mm, and the grating depths $h = 4$ and 6 mm. The 12.5 GHz surface wave is expected to be coupled to the $h = 4$ mm gratings, while the 9.5 GHz surface wave is expected to be coupled to the $h = 6$ mm gratings. Note, however, that the wave at 9.5 GHz is also coupled to the gratings for $h = 4$ mm, because the dispersion relation is satisfied. We choose here $h_R = 6$ mm and $h_L = 4$ mm in Fig. 9.2. The structural parameters of the disk are l =10 mm,

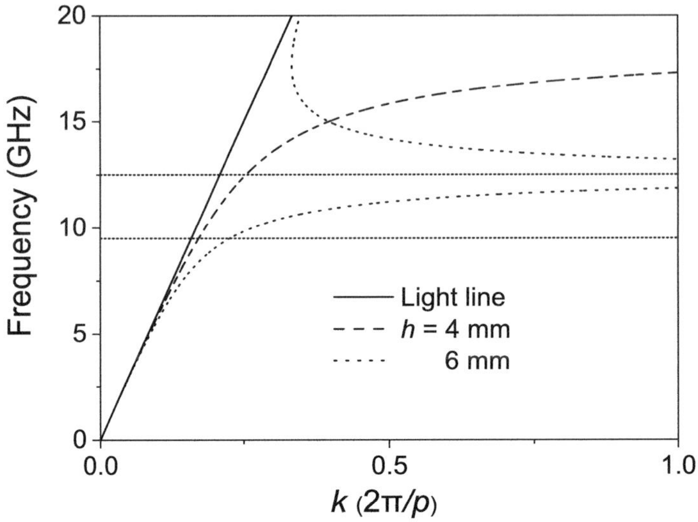

Fig. 9.5 Dispersion curves at GHz frequencies. © [2019] IEEE. Reproduced, with permission, from [23]

$r = 8\,\text{mm}$, $w_T = 16\,\text{mm}$, $w_B = 25\,\text{mm}$, $\varphi_L = \varphi_R = 135°$ and $\varphi_m = 45°$. The number of grid pairs on the surface is nine and the disk diameter is 120 mm.

We excite a linearly polarized wave in the direction of $\varphi = 0$ as the incident field. The spatial sampling grids are chosen as follows: $\Delta\rho = \Delta z = 0.2\,\text{mm}$ and $\Delta\varphi = 7.5°$. The number of sample points is 430, 300, and 48 in ρ, z, and φ directions, respectively. In addition, we impose the CPML with ten layers outside of the above computational domain. For the CPML parameters, a theoretical reflection coefficient of $R(0) = 10^{-8}$ and a quadratic conductivity profile are chosen such that the CPML provides spurious reflections below -60 dB for the present numerical example (the complex frequency-shifted technique is not adopted here).

Note that the maximum Δt_{HIE} for the present HIE-FDTD method is expressed as [12]

$$\Delta t_{\text{HIE}} \leq \frac{1}{c\sqrt{(1/\Delta\rho)^2 + (1/\rho_{\max}\Delta\varphi)^2 + (1/\Delta z)^2}} \tag{9.13}$$

where c is the speed of light in a vacuum, $\Delta\rho$, $\Delta\varphi$ and Δz are respectively the mesh sizes in ρ, φ and z directions, and $\rho_{\max}$ is the outermost value in the ρ direction. We define the time step size as CFLN = $\Delta t_{\text{HIF}}/\Delta t_{\text{CFL}}$, where Δt_{CFL} is the upper bound of the explicit cylindrical FDTD method. The maximum CFLN is calculated to be $\simeq$10.5 for the above problem. Since the field is observed to blow up with CFLN$>$10.5, the maximum CFLN is chosen to be 10 in the following.

Figure 9.6 shows the field distributions of the H_φ component on the disk surface. For the HIE-FDTD method, CFLN = 10 is used. The CPML imposed outside the computational domain is seen to work well. As predicted, the wave at 9.5 GHz is

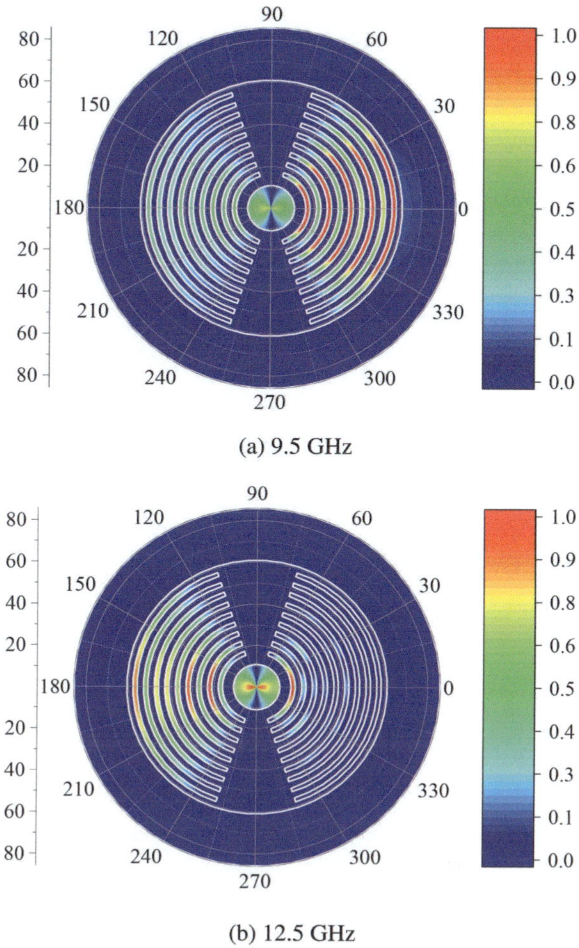

Fig. 9.6 H_ϕ components on the metal surface. © [2019] IEEE. Reproduced, with permission, from [23]

coupled to both gratings due to the fulfillment of the dispersion relation for $h_R = 6$ mm and $h_L = 4$ mm. In contrast, the wave at 12.5 GHz is only strongly coupled to the left grating. These results are consistent with the discussion of the dispersion relation in Fig. 9.5. Note that even with CFLN = 10, although not shown, both figures from HIE are indistinguishable from those from the explicit FDTD method, where the largest difference in field amplitude is less than 10^{-5}. It is worth mentioning that the computation time is reduced to 1.9 h with CFLN = 10, while that of the explicit cylindrical FDTD method is 10.5 h.

To study the function of the splitter, the radiation pattern in the ρ-z plane is calculated as shown in Fig. 9.7. Although not shown, the results of the HIE-FDTD method are perfectly superimposed on the explicit counterparts. At 9.5 GHz the radiation is observed in both directions along the gratings, i.e. $\theta = 90°$ and $270°$. This is because the wave at 9.5 GHz is coupled to the right and left gratings, as

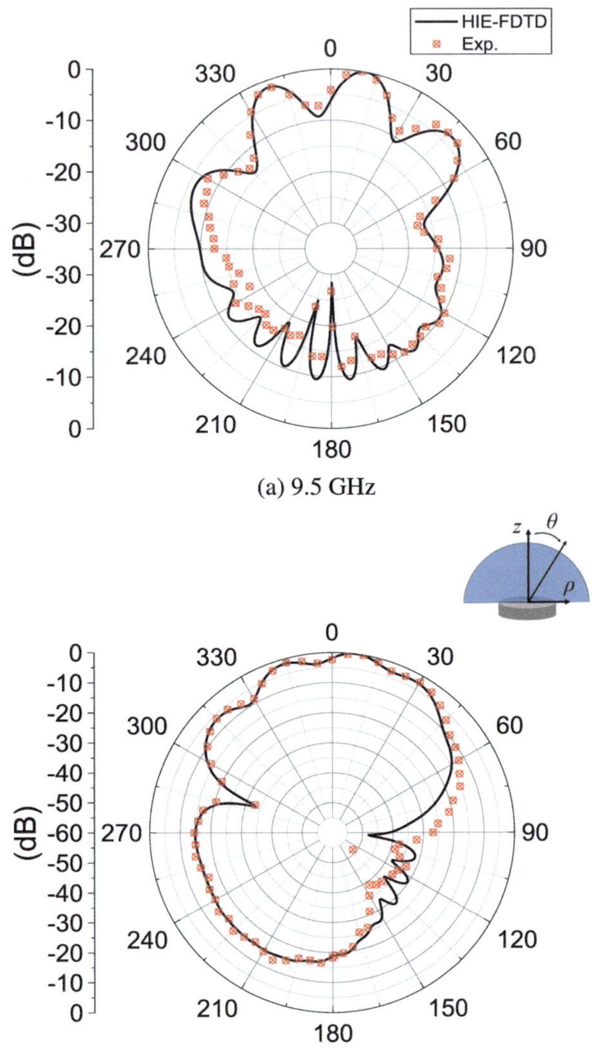

Fig. 9.7 H_ϕ components on the metal surface. © [2019] IEEE. Reproduced, with permission, from [23]

discussed in Figs. 9.5 and 10.6. In contrast, at 12.5 GHz the radiation is emitted in the left direction ($\theta = 270°$) and not in the right direction ($\theta = 90°$).

To verify the numerical results, we compare them with the experimental results. Figure 9.8 shows a fabricated metal disk splitter. The experimental results are also shown in Fig. 9.7. It is noteworthy that the experimental results are in good agreement with the numerical results, demonstrating the validity of the analysis using the cylindrical HIE-FDTD method.

Fig. 9.8 Fabricated disk splitter

9.5 Concluding Remarks

A novel terahertz surface wave splitter has been proposed by the analysis using the cylindrical FDTD method. The splitter consists of a metal disk with a central hole and radially arranged gratings. The grating depths at the right and left parts on the metal surface are designed using a one-dimensional dispersion relation. Numerical analysis shows that the surface wave at 1.0 THz propagates on the right gratings and that at 1.5 THz propagates on the left gratings. The propagation power is also calculated. At 1.0 THz, the radiated power to the top of the disk is 7%, the reflected power from the input port is 85%, and the power propagating to the right and left are 5 and 3%, respectively. At 1.5 THz, these are 68, 23, 1, and 8%, respectively. Reducing the radiated and reflected power is a future topic.

The cylindrical FDTD method is a time-consuming technique because the time step size is very small due to the fact that the sampling width in the circumferential direction becomes extremely small near the center axis. To alleviate this problem, we have developed the cylindrical HIE-FDTD method. The Sherman-Morrison formula is used to solve a cyclic matrix with the standard Thomas algorithm. The computation time of the HIE-FDTD method is reduced to 1.9 h, while that of the explicit counterpart is 10.5 h. The radiation pattern is calculated, and the numerical results are found to be in good agreement with the experimental results.

References

1. S.A. Maier, S.R. Andrews, L. Martín-Moreno, F.J. García-Vidal, Terahertz surface plasmon-polariton propagation and focusing on periodically corrugated metal wires. Phys. Rev. Lett. **97**(17), 176805 (2006)
2. Q. Gan, Z. Fu, Y.J. Ding, F.J. Bartoli, Bidirectional subwavelength slit splitter for THz surface plasmons. Opt. Exp. **15**(26), 18050–18055 (2007)

3. Y.J. Zhou, Q. Jiang, T.J. Cui, Bidirectional surface wave splitters excited by a cylindrical wire. Opt. Exp. **19**(6), 5260–5267 (2011)
4. Y.J. Zhou, T.J. Cui, Multidirectional surface-wave splitters. Appl. Phys. Lett. **98**(22), 221901 (2011)
5. J. Shibayama, J. Yamauchi, H. Nakano, Metal disc-type splitter with radially placed gratings for terahertz surface waves. Electron. Lett. **51**(4), 352–353 (2015)
6. B. Huang, G. Wang, Y. Jiang, W. Wang, A hybrid implicit-explicit FDTD scheme with weakly conditional stability. Microw. Opt. Technol. Lett. **39**(2), 97–101 (2003)
7. J. Chen, J. Wang, A three-dimensional semi-implicit FDTD scheme for calculation of shielding effectiveness of enclosure with thin slots. IEEE Trans. Electromagn. Compat. **49**(2), 354–360 (2007)
8. M. Unno, H. Asai, HIE-FDTD method for hybrid system with lumped elements and conductive media. IEEE Microw. Wirel. Compon. Lett. **21**(9), 453–455 (2011)
9. J. Wang, B. Zhou, L. Shi, C. Gao, B. Chen, A novel 3-D HIE-FDTD method with one-step leapfrog scheme. IEEE Trans. Microw. Theory Tech. **62**(6), 1275–1283 (2014)
10. K. Fujita, Hybrid Newmark-conformal FDTD method for multiphysics modeling of short spark gaps with curved metallic surfaces. IEEE J. Multiscale Multiphysics Comput. Tech. **2**, 66–77 (2017)
11. A. Van Londersele, D. De Zutter, D. Vande Ginste, A collocated 3-D HIE-FDTD scheme with PML. IEEE Microw. Wirel. Compon. Lett. **27**(7), 609–611 (2017)
12. J. Chen, J. Wang, C. Tian, Three-dimensional hybrid implicit-explicit finite-difference time-domain method in the cylindrical coordinate system. IET Microw. Antennas Propag. **3**(8), 1254–1261 (2009)
13. J.W. Thomas, *Numerical Partial Differential Equations: Finite Difference Methods* (Springer, Berlin, Germany, 1995)
14. J. Shibayama, R. Ando, J. Yamauchi, H. Nakano, An LOD-FDTD method for the analysis of periodic structures at normal incidence. IEEE Antennas Wirel. Propag. Lett. **8**, 890–893 (2009)
15. J.A. Roden, S.D. Gedney, Convolution PML (CPML): an efficient FDTD implementation of the CFS-PML for arbitrary media. Microw. Opt. Technol. Lett. **27**(5), 334–339 (2000)
16. Y. Ohtera, H. Hirose, H. Yamada, Performance analysis of CPML for the compact 2-D FDTD method in cylindrical coordinate system. IEICE Trans. Electron. **E97-C**(7), 653–660 (2014)
17. J. Shibayama, M. Ito, J. Yamauchi, H. Nakano, Analysis of spoof surface plasmon polaritons in a metal disc-type splitter using the cylindrical HIE-FDTD method, in *IEEE International Conference on Numerical Electromagnetic and Multiphysics Modeling and Optimization*, Reykjavik, Iceland (2018)
18. F.J. Garcia-Vidal, L. Martín-Moreno, J.B. Pendry, Surfaces with holes in them: new plasmonic metamaterials. J. Opt. A Pure Appl. Opt. **7**(2), S97–S101 (2005)
19. Q. Gan, Z. Fu, Y.J. Ding, F.J. Bartoli, Ultrawide-bandwidth slow-light system based on THz plasmonic graded metallic grating structures. Phys. Rev. Lett. **100**(25), 256803 (2008)
20. J. Shibayama, Y. Uchizono, S. Ozaki, J. Yamauchi, H. Nakano, Treatment of metal for the FDTD analysis of terahertz devices. Opt. Quantum Electron. **46**(2), 345–356 (2014)
21. K. Fujita, W. Odajima, FDTD simulations of THz spoof surface plasmon in metallic grating structures. IEICE Tec. Rep., EST2013-5, pp. 21–26 (2013) (in Japanese)
22. H. Amarloo, M. Neshat, S. Safavi-Naeini, Terahertz plasmonic-like waveguide for integrated sensing applications. Electron. Lett. **49**(13), 836–838 (2013)
23. J. Shibayama, T. Hara, M. Ito, J. Yamauchi, H. Nakano, Improved cylindrical HIE-FDTD method and its application to a metal disc-type surface wave splitter. IEEE Microw. Wirel. Compon. Lett. **29**(3), 177–179 (2019)

Chapter 10
Extended FDTD Algorithm for Arbitrarily-Shaped Dielectric and Dispersive Media

Abstract This chapter presents an extended FDTD algorithm for the analysis of arbitrarily-shaped dielectric and dispersive media. The conventional FDTD algorithm is formulated using Yee's mesh, which has difficulty of modeling arbitrarily-shaped media such as rods and spheres. When analyzing these structures, we need to use small spatial sampling grids, which results in long computation time, particularly for three-dimensional (3-D) problems. To alleviate this problem, we extend the contour-path effective-permittivity (CP-EP) algorithm to a 3-D problem. The CP-EP algorithm has the advantage that Yee's mesh is used as it is, and a computational cell partially filled with dielectric media can be considered with a relatively large grid. The formulation is given for fourteen cases of the partially filled cells with curved interfaces. A dielectric sphere is analyzed to demonstrate the effectiveness of the CP-EP algorithm relative to the conventional staircase approximation. We also consider 3-D dispersive media. The dispersive contour pass (DCP) algorithm, which was first applied to the Z transform technique in two-dimensional (2-D) problems, is extended to the FDTD method along with the trapezoidal recursive convolution (TRC) technique. We formulate the 3-D DCP-FDTD method based on the TRC technique for the analysis of the Drude-Lorentz model. Through the analysis of a metallic sphere, the results are found to be close to the Mie solution, rather than the staircase approximation. In addition, the staircase approximation shows spurious filed along the metal surface, while the DCP algorithm provides field distributions with suppressed spurious field.

10.1 Introduction

The FDTD method discretizes Maxwell's equations in time and space using a simple finite-difference scheme that is relatively easy to formulate and implement. The FDTD method allows electromagnetic field analysis with dielectrics of arbitrary shape. However, when performing the analysis using the conventional FDTD method, each cell is assigned a discrete value along the Cartesian coordinate system, so that the boundaries of the structure are modeled as a staircase. As a result, when the boundary of the structure contains curved surfaces or slopes, it is necessary to select a small spatial mesh for accurate analysis, resulting in a long computation time.

J. Shibayama, *FDTD Analysis of Plasmonic Devices*, Springer Series in Optical Sciences 254, https://doi.org/10.1007/978-981-96-8130-3_10

Therefore, various methods have been studied to analyze structures with curved boundaries with high accuracy [1–6]. Mohammadi et al. proposed the Contour-Path Effective-Permittivity (CP-EP) method, which can accurately handle arbitrary dielectric boundaries by deriving the effective permittivity using the contour-path algorithm [2]. The CP-EP algorithm can be implemented by slightly modifying the conventional FDTD algorithm. In addition, the Dispersive Contour-Path (DCP) algorithm using the CP-EP algorithm has been introduced into the Z-transformation-based FDTD method to treat dispersive media with high accuracy [3, 4]. However, these investigations have been limited to the application to two-dimensional (2-D) FDTD methods and have not been applied to three-dimensional (3-D) problems.

Here, we first extend the 2-D CP-EP method, which can accurately analyze dielectrics of arbitrary shapes including curved surfaces, to 3-D problems [7]. First, the 3-D formulation of the CP-EP FDTD method is given. A dielectric sphere is taken as an example of a structure containing curved surfaces, and the effective dielectric constants for each cell are derived. The dielectric sphere is analyzed and the scattering cross section is evaluated. The numerical results are compared with those of the analytical Mie solution and the conventional staircase approximation to demonstrate that the 3-D CP-EP FDTD method provides a highly accurate analysis.

We also extend the DCP-FDTD method to a 3-D version with the help of the TRC technique. The Drude-Lorentz model is adopted to handle a wider range of wavelengths. Through the analysis of a metallic sphere, it is shown that the DCP-FDTD method provides faster convergence of the numerical results relative to the staircase approximation.

10.2 3-D CP-EP Algorithm

10.2.1 Extension to 3-D Problems

We formulate an extension of the FDTD method based on the CP-EP method for 3-D problems. Figure 10.1 shows an example where the dielectric is included in the computational cell. The green and red dashed lines represent the y-z and x-z cross sections, respectively, including the cell center. Here, h, d and f represent the fill factor of the dielectric in the cell.

First, consider the following Ampere and Faraday laws to derive the effective dielectric constant for each cell.

$$\frac{\partial}{\partial t}\iint \boldsymbol{D}\cdot\boldsymbol{n}\,dS = \oint \boldsymbol{H}\cdot d\boldsymbol{l} \tag{10.1}$$

$$-\frac{\partial}{\partial t}\iint \boldsymbol{B}\cdot\boldsymbol{n}\,dS = \oint \boldsymbol{E}\cdot d\boldsymbol{l} \tag{10.2}$$

We perform the surface integral of the left side of (10.1) for the cells enclosed by the green dashed line as

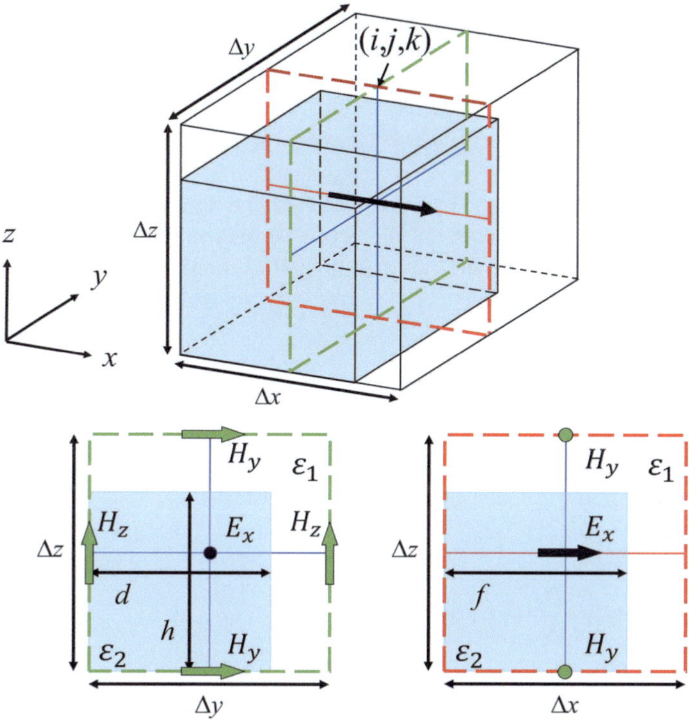

Fig. 10.1 Example of a dielectric in the cell. From [7] Copyright ©2023 by John Wiley Sons, Inc. Reproduced by permission of John Wiley & Sons, Inc.

$$
\begin{aligned}
\int_{(k-1)\Delta z}^{k\Delta z}\int_{(j-\frac{1}{2})\Delta y}^{(j+\frac{1}{2})\Delta y} D_x|_{i,y,z}dydz &= \int_{(k-1)\Delta z}^{(k-1)\Delta z+h}\left\{\int_{(j-\frac{1}{2})\Delta y}^{(j-\frac{1}{2})\Delta y+d} D_x|_{i,y,z}dy\right.\\
&\quad\left.+\int_{(j-\frac{1}{2})\Delta y+d}^{(j+\frac{1}{2})\Delta y} D_x|_{i,y,z}dy\right\}dz\\
&\quad+\int_{(k-1)\Delta z+h}^{k\Delta z}\left\{\int_{(j-\frac{1}{2})\Delta y}^{(j+\frac{1}{2})\Delta y} D_x|_{i,y,z}dy\right\}dz\\
&=\left(\frac{hd}{\Delta y\Delta z}\right)\varepsilon_2E_2+\left(1-\frac{hd}{\Delta y\Delta z}\right)\varepsilon_1E_1
\end{aligned}
\tag{10.3}
$$

where ε_1 is the relative permittivity of the background, ε_2 is the relative permittivity of the dielectric, and E_1 and E_2 are the electric fields in the x direction within each medium. The electric field component is continuous in the tangential direction, then we obtain

$$
E_1 = E_2 = E_x|_{i,j,k-\frac{1}{2}} \tag{10.4}
$$

Applying (10.4) to (10.3), the effective dielectric constant $\varepsilon_{\parallel}$ of the cell surrounded by the green dashed line can be expressed as follows:

$$\varepsilon_{\parallel} = \left(\frac{hd}{\Delta y \Delta z}\right)\varepsilon_2 + \left(1 - \frac{hd}{\Delta y \Delta z}\right)\varepsilon_1 \tag{10.5}$$

Note that in the cell enclosed by the red dashed line, the line integral of the electric field component (E_x) intersects the dielectric boundary. If the line integral on the right side of (10.2) is performed only for the path through the center of the cell, the following equation is obtained:

$$\begin{aligned}
&\int_{(i-\frac{1}{2})\Delta x}^{(i+\frac{1}{2})\Delta x} E_x|_{x,j,k-\frac{1}{2}} dx \\
&= \int_{(i-\frac{1}{2})\Delta x}^{(i-\frac{1}{2})\Delta x+f} E_x|_{x,j,k-\frac{1}{2}} dx + \int_{(i-\frac{1}{2})\Delta x+f}^{(i+\frac{1}{2})\Delta x} E_x|_{x,j,k-\frac{1}{2}} dx \\
&= \frac{f}{\Delta x} E_2 + \left(1 - \frac{f}{\Delta x}\right) E_1 \\
&= \frac{f}{\Delta x} \frac{D_2}{\varepsilon_2} + \left(1 - \frac{f}{\Delta x}\right) \frac{D_1}{\varepsilon_1}
\end{aligned} \tag{10.6}$$

The flux density is continuous in the vertical direction, the we obtain

$$D_1 = D_2 = D_x|_{i,j,k-\frac{1}{2}} \tag{10.7}$$

Applying (10.6) into (10.7) results in

$$\begin{aligned}
&\left\{ \frac{\left(\frac{f}{\Delta x}\right)}{\varepsilon_2} + \frac{\left(1 - \frac{f}{\Delta x}\right)}{\varepsilon_1} \right\} D_x|_{i,j,k-\frac{1}{2}} \\
&= \frac{1}{\varepsilon_\perp} D_x|_{i,j,k-\frac{1}{2}} = E'_x|_{i,j,k-\frac{1}{2}}
\end{aligned} \tag{10.8}$$

where $\varepsilon_\perp$ and $E'_x|_{i,j,k-\frac{1}{2}}$ represent the effective dielectric constant and the electric field component, taking into account the fact that the line integral of the electric field component crosses the dielectric boundary.

From the above, considering both the Ampare's area integral and the Faraday's line integral, we can express the flux density as

$$\varepsilon_\parallel E_x|_{i,j,k-\frac{1}{2}} = \varepsilon_\parallel \frac{D_x|_{i,j,k-\frac{1}{2}}}{\varepsilon_2} = \varepsilon_\parallel \frac{\varepsilon_\perp}{\varepsilon_2} E'_x|_{i,j,k-\frac{1}{2}} \tag{10.9}$$

Then, the effective dielectric constant of the cell shown in Fig. 10.1 is derived as

$$\varepsilon_{\text{eff}} = \frac{\varepsilon_\parallel \varepsilon_\perp}{\varepsilon_2} \tag{10.10}$$

By incorporating the effective dielectric constant derived from (10.10) into the ordinary FDTD method for the cells in Fig. 10.1, it is possible to accurately analyze the case where a dielectric boundary is included in a cell, even in three dimensions. The same procedure can be used to derive the effective dielectric constant for cells for which E_y and E_z are calculated.

10.2.2 Application to the Curved Surface

The CP-EP method can also be applied to structures with curved surfaces using the method formulated in the previous section. Figure 10.2 shows the model of the dielectric sphere to be analyzed. In the 2-D model of the literature [2], six dielectric configurations must be considered when analyzing a cylindrical dielectric. On the other hand, the 3-D dielectric sphere model shown in Fig. 10.2 requires the derivation of the effective dielectric constant considering 14 dielectric configurations.

Figure 10.3 shows an example where the dielectric boundary in the computational cell intersects Faraday's integration line (red line) and the dielectric contains the center of the computational cell, where $\boldsymbol{p}$, $\boldsymbol{q}$, $\boldsymbol{r}$ is the unit normal vector in the x, y and z directions. In the case of Fig. 10.3, since the dielectric boundary in the cell and the integration line (red line) intersect, it is necessary to consider the effective dielectric constant $\varepsilon_{\perp}$ by line integration of Faraday's law as shown below.

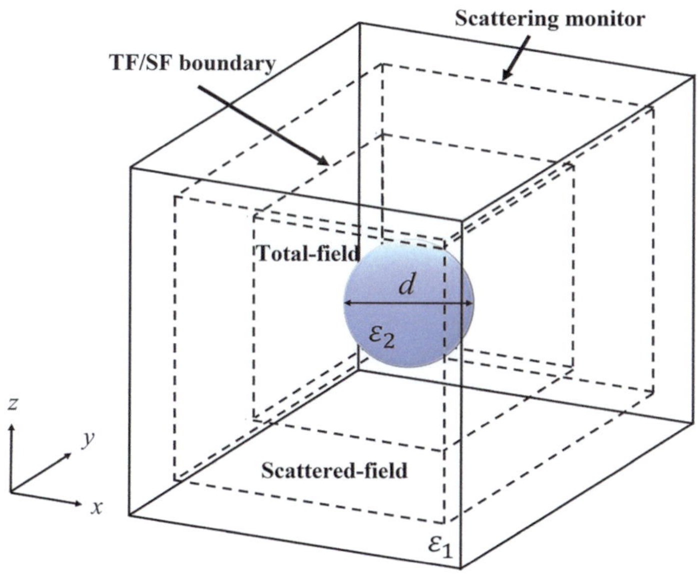

Fig. 10.2 Calculation model of a dielectric sphere

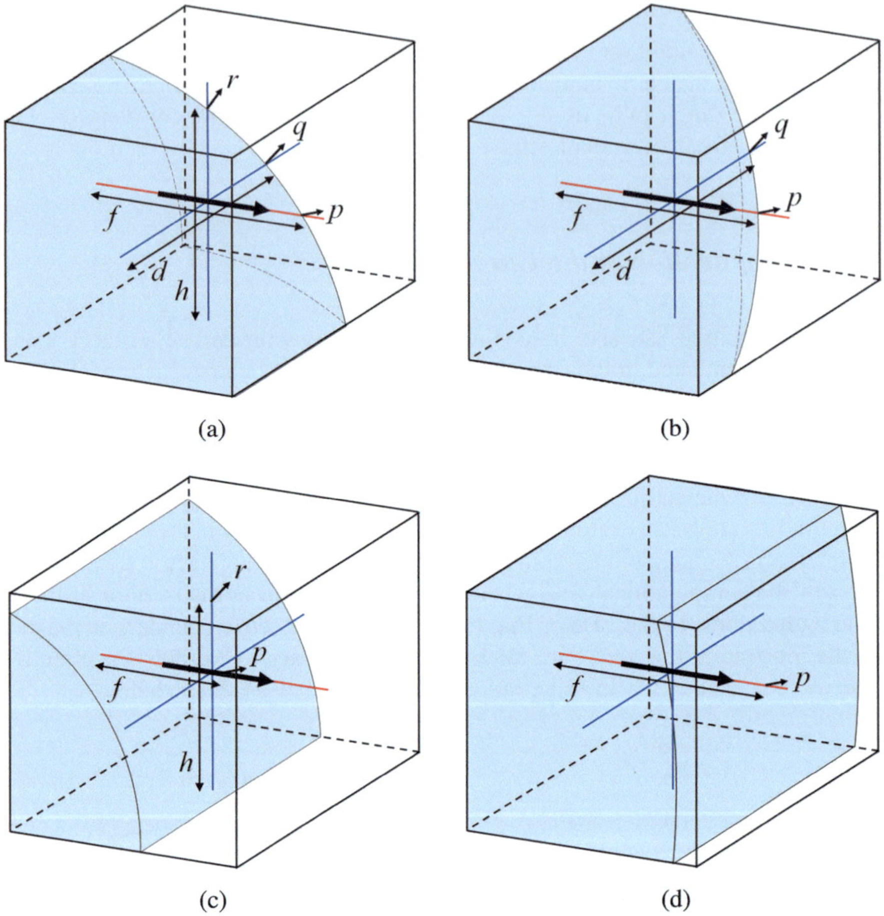

Fig. 10.3 Partially filled cells with curved interfaces, where the integral line (red line) for the Faraday law crosses the dielectric/background interface and the center of the computational cell is in the dielectric material. From [7] Copyright ©2023 by John Wiley Sons, Inc. Reproduced by permission of John Wiley & Sons, Inc.

$$\varepsilon_{\perp,p} = \left\{ \frac{\frac{f}{\Delta x}}{\varepsilon_2} + \left(1 - \frac{f}{\Delta x}\right)\left(\frac{p^2}{\varepsilon_1} + \frac{1 - p^2}{\varepsilon_2}\right) \right\}^{-1} \tag{10.11}$$

where p is the projection of the unit normal vector onto the electric field components. Next, considering the intersection of the dielectric boundary in the cell and the integration line (blue line) according to Ampere's law, the effective dielectric constants $\varepsilon_{\parallel}$ in cases Fig. 10.3a, b, c, and d can be expressed by the following equations, respectively:

$$\varepsilon_{\parallel,r,q} = \varepsilon_2 \frac{dh}{\Delta y \Delta z} + (\varepsilon_2 q^2 + \varepsilon_1(1-q^2))\Big(1 - \frac{dh}{\Delta y \Delta z}\Big) \tag{10.12a}$$

$$\varepsilon_{\parallel,q} = \varepsilon_2 \frac{d}{\Delta y} + (\varepsilon_2 q^2 + \varepsilon_1(1-q^2))\Big(1 - \frac{d}{\Delta y}\Big) \tag{10.12b}$$

$$\varepsilon_{\parallel,r} = \varepsilon_2 \frac{h}{\Delta z} + (\varepsilon_2 r^2 + \varepsilon_1(1-r^2))\Big(1 - \frac{h}{\Delta z}\Big) \tag{10.12c}$$

$$\varepsilon_{\parallel} = \varepsilon_2 \tag{10.12d}$$

where q and r are the projections of the unit normal vector onto the electric field components, and in case (a), $p = r$. From the above, it follows that the effective dielectric constant for the cell shown in Fig. 10.3 can be derived by applying the coefficients of (10.11) and (10.12a)–(10.12d) to (10.10).

Next, Fig. 10.4 shows an example where the dielectric boundary in the calculation cell intersects Faraday's line integral (red line) and the dielectric does not include the center of the calculation cell. In Fig. 10.4, since the dielectric does not include the center of the cell, the effective dielectric constant $\varepsilon_{\perp}$ based on the line integral of Faraday's law can be expressed by the following equation:

$$\varepsilon_{\perp,p} = \left\{ \frac{f}{\Delta x}\Big(\frac{p^2}{\varepsilon_2} + \frac{1-p^2}{\varepsilon_1}\Big) + \frac{1-\frac{f}{\Delta x}}{\varepsilon_1} \right\}^{-1} \tag{10.13}$$

Using the same technique, the effective dielectric constant $\varepsilon_{\parallel}$ by integration of Ampere's law in the cases Fig. 10.4a, b, c, and d can be derived as follows:

$$\varepsilon_{\parallel,r,q} = (\varepsilon_1 q^2 + \varepsilon_2(1-q^2)) \frac{dh}{\Delta y \Delta z} + \varepsilon_1\Big(1 - \frac{dh}{\Delta y \Delta z}\Big) \tag{10.14a}$$

$$\varepsilon_{\parallel,q} = (\varepsilon_1 q^2 + \varepsilon_2(1-q^2)) \frac{d}{\Delta y} + \varepsilon_1\Big(1 - \frac{d}{\Delta y}\Big) \tag{10.14b}$$

$$\varepsilon_{\parallel,r} = (\varepsilon_1 r^2 + \varepsilon_2(1-r^2)) \frac{h}{\Delta z} + \varepsilon_1\Big(1 - \frac{h}{\Delta z}\Big) \tag{10.14c}$$

$$\varepsilon_{\parallel} = \varepsilon_1 \tag{10.14d}$$

Considering that the dielectric does not include the center of the computed cell, the effective dielectric constant for the cell shown in Fig. 10.4 can be derived by applying the coefficients of (10.13) and (10.14a)–(10.14d) to (10.15) as

$$\varepsilon_{\text{eff}} = \frac{\varepsilon_{\parallel}\varepsilon_{\perp}}{\varepsilon_1} \tag{10.15}$$

Finally, Fig. 10.5 shows an example of a case where the dielectric boundary in the cell and the Faraday's law integration line (red line) do not intersect. In the case of Fig. 10.5, since the dielectric boundary in the cell and the integration line (red line) do not intersect, only the effective dielectric constant $\varepsilon_{\parallel}$ due to integration of Ampere's law is considered, as shown in (10.16) below.

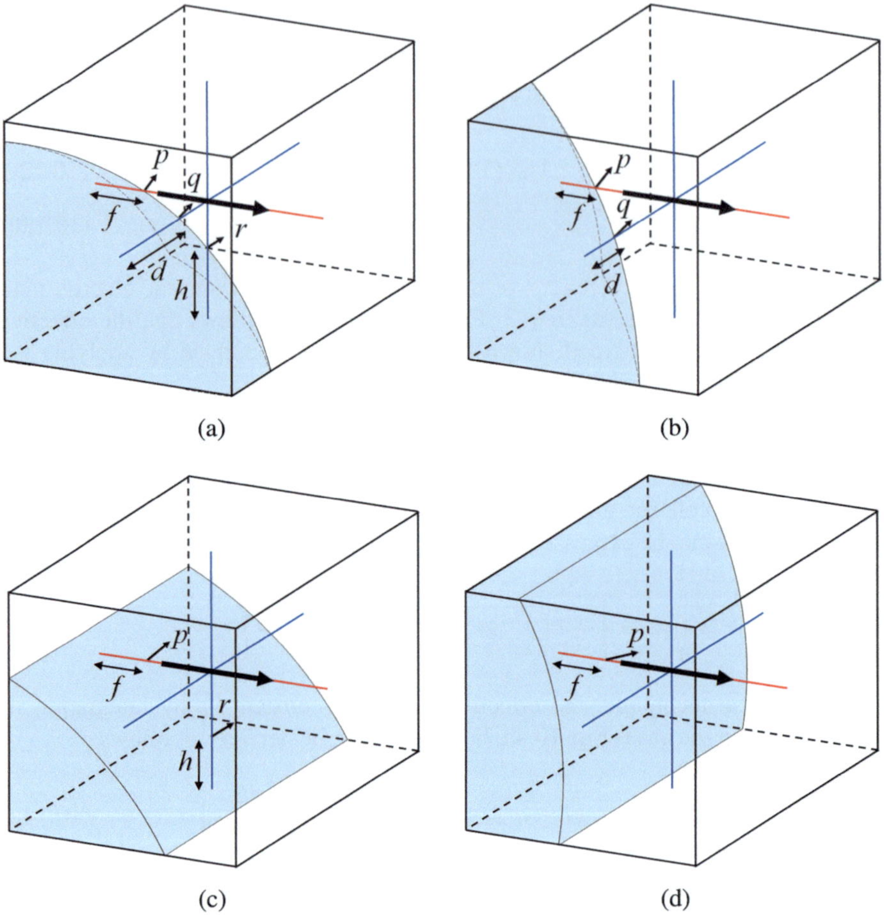

Fig. 10.4 Partially filled cells with curved interfaces, where the integral line (red line) for the Faraday law crosses the dielectric/background interface and the center of the computational cell is outside of the dielectric material. From [7] Copyright ©2023 by John Wiley Sons, Inc. Reproduced by permission of John Wiley & Sons, Inc.

$$\varepsilon_{\text{eff}} = \frac{\varepsilon_{\parallel}\varepsilon_{\perp}}{\varepsilon} = \varepsilon_{\parallel} \tag{10.16}$$

As in Figs. 10.3 and 10.4, the effective dielectric constants for the cells in (a)–(f) in Fig. 10.5 can be derived by applying the coefficients in (10.12a)–(10.12b) and (10.14a)–(10.14b) to each dielectric configuration. The effective dielectric constant can also be derived for dielectrics of arbitrary shapes other than spheres by the same method as above.

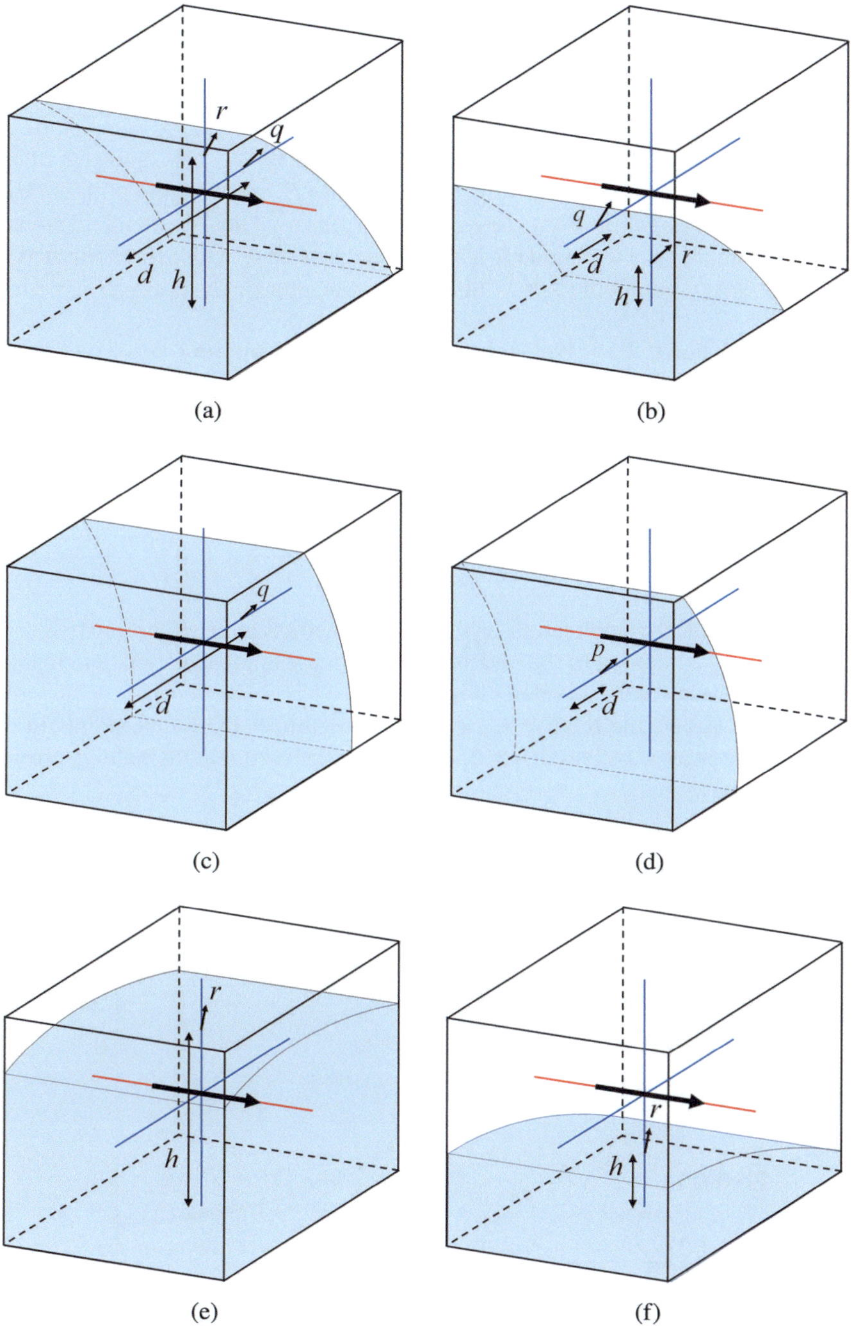

Fig. 10.5 Partially filled cells with curved interfaces, where the integral line (red line) for the Faraday law doesn't cross the dielectric/background interface. From [7] Copyright ©2023 by John Wiley Sons, Inc. Reproduced by permission of John Wiley & Sons, Inc.

10.2.3 Analysis of Dielectric Sphere

The CP-EP method is used to analyze the dielectric sphere model shown in Fig. 10.2. First, the scattering cross section (SCS) is evaluated. The dielectric constant of the sphere is set to $\varepsilon_2 = 5$ and the background is air ($\varepsilon_1 = 1$). The diameter of the dielectric sphere is set to $d = 240\,\text{nm}$. We excite the E_x polarization with uniform amplitude in the x direction along the $+z$ axis using the total field/scatter field domain decomposition (TF/SF) technique [8]. The spatial sampling width is determined by the number of sampling points N with a sphere diameter d, i.e. $\Delta x = \Delta y = \Delta z = d/N$.

Figure 10.6 shows the wavelength response of the scattering cross section for $N = 20$. For comparison, the results from the staircase approximation and the results from Mie theory [9] are also shown. In order to compare the calculated and Mie theory results, the relative error is evaluated in Fig. 10.7. The relative error is expressed by the following equation:

$$\text{Relative error}[\%] = \frac{|\text{SCS}_{\text{FDTD}} - \text{SCS}_{\text{Mie}}|}{\text{SCS}_{\text{Mie}}} \times 100 \tag{10.17}$$

It can be seen that the application of the CP-EP method reduces the relative error with Mie theory compared to the results of the staircase approximation, and that the variation of the error with wavelength is small.

Figure 10.8 shows the relative error versus the number of sample points in the diameter at the peak wavelength λ = 558 nm. It can be seen that the relative error to

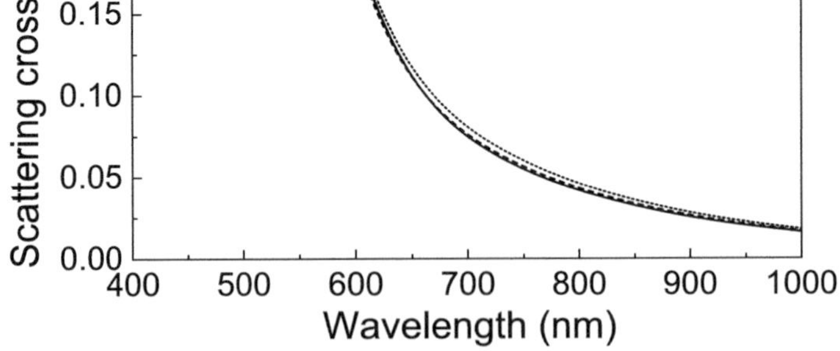

Fig. 10.6 Scattering cross-section versus wavelength. From [7] Copyright ©2023 by John Wiley Sons, Inc. Reproduced by permission of John Wiley & Sons, Inc.

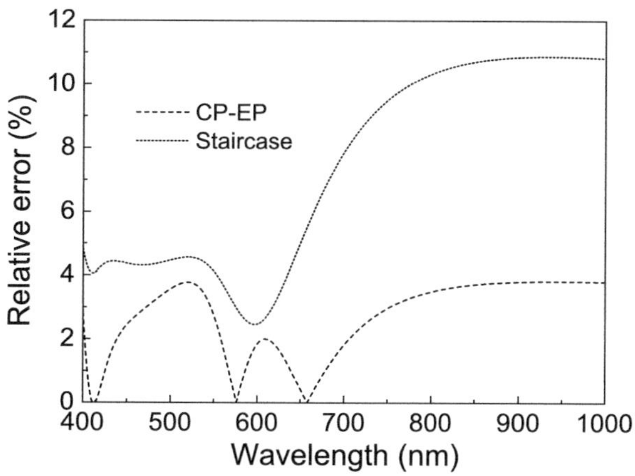

Fig. 10.7 Relative error versus wavelength. From [7] Copyright ©2023 by John Wiley Sons, Inc. Reproduced by permission of John Wiley & Sons, Inc.

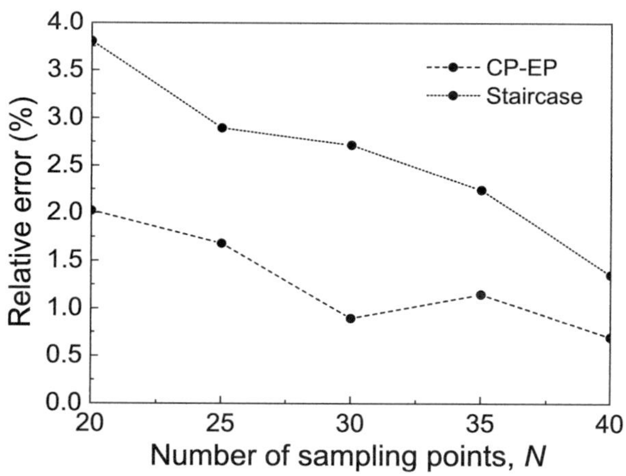

Fig. 10.8 Relative error as a function of the number of sampling points. From [7] Copyright ©2023 by John Wiley Sons, Inc. Reproduced by permission of John Wiley & Sons, Inc.

Mie theory is about half in the case of the CP-EP method compared to the case using the staircase approximation. The CP-EP method is more accurate than the staircase approximation even for relatively large sampling widths.

Finally, the field distribution (E_x component) near the dielectric sphere is evaluated. The wavelength is chosen to be $\lambda = 558$ nm and the number of sphere diameter divisions is $N = 40$. Figure 10.9a shows the field distribution when the CP-EP method is used, and (b) shows the field distribution when the staircase approximation is used. It can be seen that the CP-EP method produces a smooth field along the sphere at the boundary between the dielectric and the air. On the other hand, when the staircase approximation is used, the field distribution at the boundary is distorted.

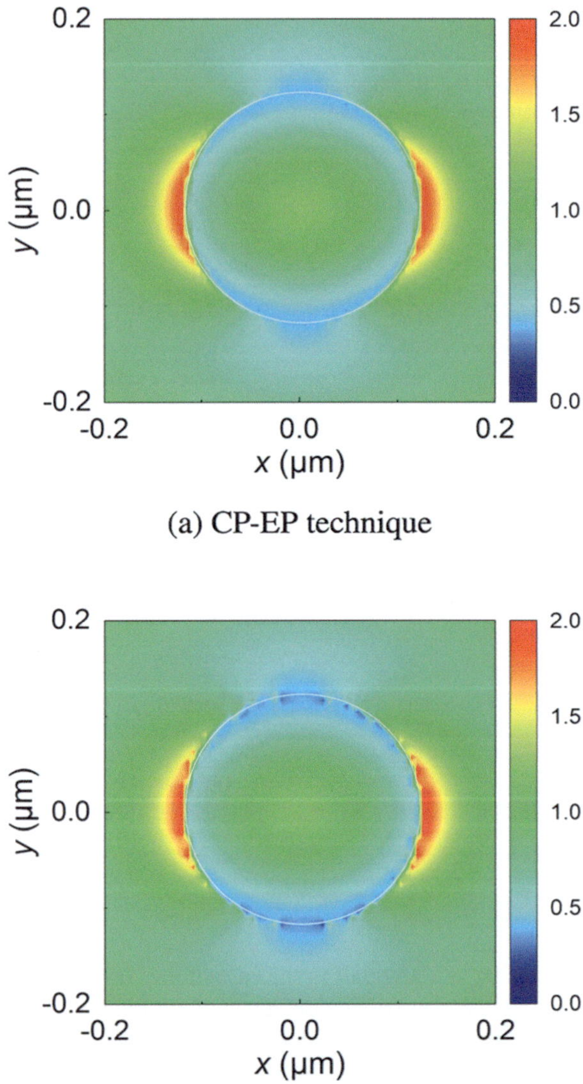

(a) CP-EP technique

(b) Staircase

Fig. 10.9 Field distributions. From [7] Copyright ©2023 by John Wiley Sons, Inc. Reproduced by permission of John Wiley & Sons, Inc.

10.3 3-D DCP Algorithm

10.3.1 Extension to 3-D Problems

We here extend the DCP method, which can accurately analyze dispersive media of arbitrary shapes including curved surfaces, to 3-D problems. First, the Drude-Lorentz model is incorporated into the FDTD method based on the TRC technique. In the frequency domain, the relative permittivity in a dispersive medium is expressed by the following Drude-Lorentz model:

$$\varepsilon_r(\omega) = \varepsilon_\infty + \frac{\omega_{\mathrm{D}}^2}{j\omega(\nu_{\mathrm{D}} + j\omega)} + \frac{\Delta\varepsilon\omega_{\mathrm{L}}^2}{j\omega\nu_{\mathrm{L}} + \omega_{\mathrm{L}}^2 - \omega^2} \tag{10.18}$$

where ε_∞ is the dielectric constant of the material at infinite frequency, ω is the angular frequency, ω_{D} and ω_p are the electron plasma frequencies, ν_{D} and ν_p are the effective electron collision frequencies, and $\Delta\varepsilon$ is the weighting coefficient. Compared to (3.1), only one pole is retained for the Lorentz function. The flux density in a dispersive medium can be expressed by the following equation using the TRC technique [10]:

$$\begin{aligned} \boldsymbol{D}^n &= \varepsilon_\infty \boldsymbol{E}^n + \frac{\boldsymbol{E}^n + \boldsymbol{E}^{n-1}}{2}\chi^0 + \sum_{m=0}^{n-2} \frac{\boldsymbol{E}^{n-m-1} + \boldsymbol{E}^{n-m-2}}{2}\chi^{m+1} \\ &\equiv \varepsilon_\infty(\boldsymbol{E}^n + \boldsymbol{\eta}^n) \end{aligned} \tag{10.19}$$

where χ is the electric susceptibility. Substituting (10.19) into Ampere-Maxwell law, we obtain the update equation for the electric field as

$$\begin{aligned} \boldsymbol{E}^n &= \frac{1 - C_a}{1 + C_a}\boldsymbol{E}^{n-1} + \frac{D_a c\Delta t}{1 + C_a}\nabla \times \boldsymbol{H}^{n-\frac{1}{2}} \\ &+ \frac{1 - C_b}{1 + C_a}\boldsymbol{\eta}^{n-1} + \frac{1}{1 + C_a}\boldsymbol{\phi}^{n-1} \end{aligned} \tag{10.20}$$

where c is the speed of light in a vacuum, Δt is the time step size. Each variable in (10.20) is expressed as

$$\begin{aligned} \boldsymbol{\eta}^n &= C_a(\boldsymbol{E}^n + \boldsymbol{E}^{n-1}) - \boldsymbol{\phi}^{n-1} + C_b\boldsymbol{\eta}^{n-1} \\ \boldsymbol{\phi}^n &= \boldsymbol{\phi}_{\mathrm{D}}^n + \mathrm{Re}(\boldsymbol{\phi}_{\mathrm{L}}^{\mathrm{n}}) \\ \boldsymbol{\phi}_{\mathrm{D}}^n &= C_c(\boldsymbol{E}^n + \boldsymbol{E}^{n-1}) + e^{-\nu_{\mathrm{D}}\Delta t}\boldsymbol{\phi}_{\mathrm{D}}^{n-1} + C_d(\boldsymbol{\eta}^n + \boldsymbol{\eta}^{n-1}) \\ \boldsymbol{\phi}_{\mathrm{L}}^n &= C_e(\boldsymbol{E}^n + \boldsymbol{E}^{n-1}) + e^{\gamma\Delta t}\boldsymbol{\phi}_{\mathrm{L}}^{n-1} + C_f(\boldsymbol{\eta}^n + \boldsymbol{\eta}^{n-1}) \end{aligned} \tag{10.21}$$

where $\gamma = -\alpha + j\beta$, $\alpha = \nu_{\mathrm{L}}/2$ and $\beta = \sqrt{(\omega_{\mathrm{L}}^2 - \alpha^2)}$. Each coefficient in the dispersive material is

$$\begin{aligned} &D_a = \frac{1}{\varepsilon_\infty},\ C_a = \frac{\chi^0}{2\varepsilon_\infty},\ C_b = 1,\ C_c = \frac{\Delta\chi_{\mathrm{D}}^0}{2\varepsilon_\infty}, \\ &C_d = 0,\ C_e = \frac{\Delta\chi_{\mathrm{L}}^0}{2\varepsilon_\infty},\ C_f = 0 \end{aligned} \tag{10.22}$$

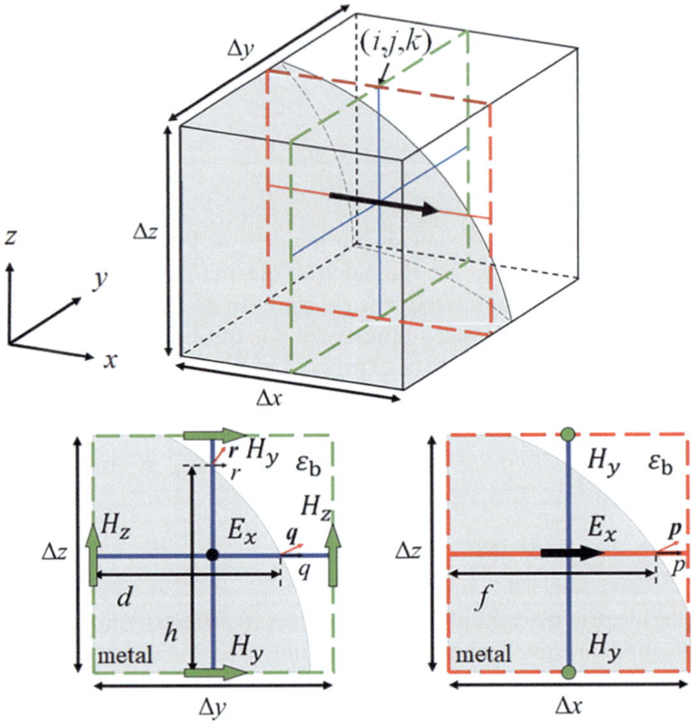

Fig. 10.10 Example of a dielectric in the cell

where $\chi^0 = \chi_D^0 + \mathrm{Re}(\chi_L^0)$. The coefficients of $C_a \sim C_f$ and D_a are derived later.

We derive the coefficients $C_a \sim C_f$ and D_a, which are determined by the medium configuration in the computational cell. Figure 10.10 shows an example of a calculation cell containing a metal. The green and red dashed lines represent y-z and x-z cross sections including the cell center, respectively, where f, d, and h are the fill factor of the medium in the cell, $\boldsymbol{p}$, $\boldsymbol{q}$, and $\boldsymbol{r}$ are the unit normal vectors in the x, y, and z directions, and p, q, r are the projections of the unit normal vectors to the electric field components G.

First, we consider the following Ampere and Faraday's law to derive the coefficients for each cell:

$$\frac{\partial}{\partial t}\iint \boldsymbol{D}\cdot\boldsymbol{n}\,dS = \oint \boldsymbol{H}\cdot d\boldsymbol{l} \tag{10.23}$$

$$-\frac{\partial}{\partial t}\iint \boldsymbol{B}\cdot\boldsymbol{n}\,dS = \oint \boldsymbol{E}\cdot d\boldsymbol{l} \tag{10.24}$$

For the cell enclosed by the green dashed line, the following equation is obtained by performing the area integral on the left side of (10.23):

$$\begin{aligned}\int_{(k-1)\Delta z}^{k\Delta z}\int_{(j-\frac{1}{2})\Delta y}^{(j+\frac{1}{2})\Delta y} D_x|_{i,y,z}dydz &= \int_{(k-1)\Delta z}^{(k-1)\Delta z+h}\left\{\int_{(j-\frac{1}{2})\Delta y}^{(j-\frac{1}{2})\Delta y+d} D_x|_{i,y,z}dy\right.\\ &\quad\left.+\int_{(j-\frac{1}{2})\Delta y+d}^{(j+\frac{1}{2})\Delta y} D_x|_{i,y,z}dy\right\}dz\\ &\quad+\int_{(k-1)\Delta z+h}^{k\Delta z}\left\{\int_{(j-\frac{1}{2})\Delta y}^{(j+\frac{1}{2})\Delta y} D_x|_{i,y,z}dy\right\}dz\\ &= \varepsilon_\infty\{hd+(1-hd)q^2\}(E_2+\eta_2)\\ &\quad+\varepsilon_b(1-hd)(1-q^2)E_1 \end{aligned} \tag{10.25}$$

where ε_b is the relative permittivity in the background, ε_∞ is the relative permittivity of the dispersive material at an infinity frequency, and E_1 and E_2 are the electric field in the x direction for each material. The tangential component of the electric field is continuous, so that

$$E_1 = E_2 = E_x|_{i,j,k-\frac{1}{2}} \tag{10.26}$$

Substituting (10.26) to (10.25) leads to

$$E_2 = \frac{1}{\varepsilon_{\|,q,r}}D_x|_{i,j,k-\frac{1}{2}} - \frac{\varepsilon_\infty}{\varepsilon_{\|,q,r}}c_{\|,q,r}\eta_2 \tag{10.27}$$

where

$$\begin{aligned}\varepsilon_{\|,q,r} &= \varepsilon_\infty hd + \{\varepsilon_\infty q^2 + \varepsilon_b(1-q^2)\}(1-hd)\\ c_{\|,q,r} &= hd + (1-hd)q^2\end{aligned} \tag{10.28}$$

Note that in the cell enclosed by the red dashed line, the line integral of the electric field component (E_x) intersects the metal/background boundary. We perform line integration on the right-hand side of (10.24) only for the path through the cell center as

$$\begin{aligned}&\int_{(i-\frac{1}{2})\Delta x}^{(i+\frac{1}{2})\Delta x} E_x|_{x,j,k-\frac{1}{2}}dx\\ &= \int_{(i-\frac{1}{2})\Delta x}^{(i-\frac{1}{2})\Delta x+f} E_x|_{x,j,k-\frac{1}{2}}dx + \int_{(i-\frac{1}{2})\Delta x+f}^{(i+\frac{1}{2})\Delta x} E_x|_{x,j,k-\frac{1}{2}}dx\\ &= (1-f)\left\{\frac{p^2}{\varepsilon_b}D_1 + (1-p^2)E_2\right\} + fE_2\end{aligned} \tag{10.29}$$

The normal component of the flux density is continuous at the boundaries of different media, so that

$$D_1 = D_2 = D_x|_{i,j,k-\frac{1}{2}} \tag{10.30}$$

Substituting (10.30) to (10.29) results in

$$E'_x|_{i,j,k-\frac{1}{2}} = \frac{\varepsilon_\infty}{\varepsilon_{\perp,p}} E_2 + \frac{\varepsilon_\infty}{\varepsilon_\mathrm{b}} c_{\perp,p}\eta_2 \tag{10.31}$$

where

$$\begin{aligned} \varepsilon_{\perp,p} &= \left\{ \frac{f}{\varepsilon_\infty} + (1-f)\left(\frac{p^2}{\varepsilon_\mathrm{b}} + \frac{1-p^2}{\varepsilon_\infty} \right) \right\}^{-1} \\ c_{\perp,p} &= (1-f)p^2 \end{aligned} \tag{10.32}$$

Note in (10.31) that $E'_x|_{i,j,k-\frac{1}{2}}$ represents the electric field component considering the fact that the line integral of the electric field component crosses the metal/background boundary.

Substituting (10.27) to (10.31) yields

$$\begin{aligned} E'_x|_{i,j,k-\frac{1}{2}} &= \frac{\varepsilon_\infty}{\varepsilon_{\parallel,q,r}\varepsilon_{\perp,p}} D_x|_{i,j,k-\frac{1}{2}} - \varepsilon_\infty \left(\frac{\varepsilon_\infty c_{\parallel,q,r}}{\varepsilon_{\parallel,q,r}\varepsilon_{\perp,p}} - \frac{c_{\perp,p}}{\varepsilon_\mathrm{b}} \right)\eta_2 \\ &\equiv \frac{\varepsilon_\infty}{\varepsilon_{\parallel,q,r}\varepsilon_{\perp,p}} D_x|_{i,j,k-\frac{1}{2}} - \eta \end{aligned} \tag{10.33}$$

Substituting (10.33) to (10.21) leads to the following coefficients:

$$D_{a(a)} = \frac{\varepsilon_\infty}{\varepsilon_{\parallel,q,r}\varepsilon_{\perp,p}} \tag{10.34a}$$

$$C_{a(a)} = \frac{\chi^0}{2\varepsilon_\infty} \frac{\varepsilon_{\perp,p}\left(\frac{\varepsilon_\infty c_{\parallel,q,r}}{\varepsilon_{\parallel,q,r}\varepsilon_{\perp,p}} - \frac{c_{\perp,p}}{\varepsilon_\mathrm{b}} \right)}{1 + C_{g(a)}} \tag{10.34b}$$

$$C_{b(a)} = \frac{1 - C_{g(a)}}{1 + C_{g(a)}} \tag{10.34c}$$

$$C_{c(a)} = \frac{\Delta\chi_\mathrm{D}^0}{2\varepsilon_\infty} \frac{\varepsilon_{\perp,p}\left(\frac{\varepsilon_\infty c_{\parallel,q,r}}{\varepsilon_{\parallel,q,r}\varepsilon_{\perp,p}} - \frac{c_{\perp,p}}{\varepsilon_\mathrm{b}} \right)}{1 + C_{h(a)}} \tag{10.34d}$$

$$C_{d(a)} = -\frac{\Delta\chi_\mathrm{D}^0}{2\varepsilon_\infty} \frac{\varepsilon_{\perp,p} c_{\perp,p}}{\varepsilon_\mathrm{b}(1 + C_{h(a)})} \tag{10.34e}$$

$$C_{e(a)} = \frac{\Delta\chi_\mathrm{L}^0}{2\varepsilon_\infty} \frac{\varepsilon_{\perp,p}\left(\frac{\varepsilon_\infty c_{\parallel,q,r}}{\varepsilon_{\parallel,q,r}\varepsilon_{\perp,p}} - \frac{c_{\perp,p}}{\varepsilon_\mathrm{b}} \right)}{1 + C_{i(a)}} \tag{10.34f}$$

$$C_{f(a)} = -\frac{\Delta\chi_\mathrm{L}^0}{2\varepsilon_\infty} \frac{\varepsilon_{\perp,p} c_{\perp,p}}{\varepsilon_\mathrm{b}(1 + C_i(a))} \tag{10.34g}$$

$$C_{g(a)} = \frac{\chi^0 \varepsilon_{\perp,p} c_{\perp,p}}{2\varepsilon_\infty \varepsilon_\mathrm{b}} \tag{10.34h}$$

$$C_{h(a)} = \frac{\chi_{\mathrm{D}}^{0}\varepsilon_{\perp,p}c_{\perp,p}}{2\varepsilon_{\infty}\varepsilon_{\mathrm{b}}} \tag{10.34i}$$

$$C_{i(a)} = \frac{\chi_{\mathrm{L}}^{0}\varepsilon_{\perp,p}c_{\perp,p}}{2\varepsilon_{\infty}\varepsilon_{\mathrm{b}}} \tag{10.34j}$$

By applying these coefficients to (10.20) and (10.21), the analysis can be performed accurately even in three dimensions when the cell contains a dispersive medium boundary. The coefficients for each cell for which E_y and E_z are derived in the same way.

10.3.2 Application to the Curved Surface

The DCP algorithm can be applied in the same way to dispersive media of various arbitrary shapes using the method formulated in the previous section. As an example, we analyze a metallic sphere model, which requires the derivation of coefficients for 14 different media configurations, as in the case of the CP-EP method.

We first investigate an example of a case where the metal/background boundary intersects Faraday's line of integration (red line) in a calculation cell and the medium includes the center of the calculation cell. This corresponds to the case shown in Fig. 10.3. In this case, since the integration line (red line) intersects the medium boundary in the cell, the effective dielectric constant $\varepsilon_{\perp}$ and the coefficient $c_{\perp}$ of line integration of Faraday's law shown in (10.32) must be considered. Considering that the medium boundary in the cell and the integration line (blue line) by Ampere's law intersect, the effective dielectric constant $\varepsilon_{\parallel}$ and the coefficient $c_{\parallel}$ in the cases Fig. 10.3b, c and d can be expressed by the following equations, respectively (the equations of the case Fig. 10.3a are (10.28)):

$$\begin{aligned} \varepsilon_{\parallel,q} &= \varepsilon_{\infty}d + \{\varepsilon_{\infty}q^2 + \varepsilon_{\mathrm{b}}(1-q^2)\}(1-d) \\ c_{\parallel,q} &= d + (1-d)q^2 \end{aligned} \tag{10.35a}$$

$$\begin{aligned} \varepsilon_{\parallel,r} &= \varepsilon_{\infty}h + \{\varepsilon_{\infty}r^2 + \varepsilon_{\mathrm{b}}(1-r^2)\}(1-h) \\ c_{\parallel,r} &= h + (1-h)r^2 \end{aligned} \tag{10.35b}$$

$$\begin{aligned} \varepsilon_{\parallel} &= \varepsilon_{\infty} \\ c_{\parallel} &= 1 \end{aligned} \tag{10.35c}$$

Since the medium in Fig. 10.3 contains the center of the computation cell, the coefficients $C_a \sim C_f$ and D_a can be derived by using (10.34).

Next, we examine an example where the medium boundary in the calculation cell intersects Faraday's line integral (red line) and the medium does not include the center of the calculation cell. This corresponds to the case shown in Fig. 10.4. In this case, the effective permittivity $\varepsilon_{\perp}$ and the coefficient $g_{\perp}$ of Faraday's law can be expressed as follows because the medium does not include the center of the calculation cell:

$$\varepsilon_{\perp,p} = \left\{ f\left(\frac{p^2}{\varepsilon_\infty} + \frac{1-p^2}{\varepsilon_b}\right) + \frac{1-f}{\varepsilon_b} \right\}^{-1}$$
$$g_{\perp,p} = fp^2 \tag{10.36}$$

Using the same technique, the effective dielectric constants $\varepsilon_\parallel$ and $g_\parallel$ by integration of Ampere's law for (a), (b), (c) and (d) in Fig. 10.4 can be expressed by the following equations:

$$\varepsilon_{\parallel,q,r} = \{\varepsilon_b q^2 + \varepsilon_\infty(1-q^2)\}dh + \varepsilon_b(1-dh)$$
$$g_{\parallel,q,r} = hd(1-q^2) \tag{10.37a}$$
$$\varepsilon_{\parallel,q} = \{\varepsilon_b q^2 + \varepsilon_\infty(1-q^2)\}d + \varepsilon_b(1-d)$$
$$g_{\parallel,q} = d(1-q^2) \tag{10.37b}$$
$$\varepsilon_{\parallel,r} = \{\varepsilon_b r^2 + \varepsilon_\infty(1-r^2)\}h + \varepsilon_b(1-h),$$
$$g_{\parallel,r} = h(1-r^2) \tag{10.37c}$$
$$\varepsilon_\parallel = \varepsilon_\mathrm{b}$$
$$g_\parallel = 0 \tag{10.37d}$$

Also, considering that the medium in Fig. 10.4 does not contain the center of the computation cell, we derive the coefficients $C_a \sim C_f$ and D_a in the same way as in the previous section as

$$D_{a(b)} = \frac{\varepsilon_b}{\varepsilon_\parallel \varepsilon_\perp} \tag{10.38a}$$
$$C_{a(b)} = \frac{\chi^0}{2\varepsilon_\infty}\frac{\varepsilon_\perp}{\varepsilon_b}\left\{\frac{\varepsilon_b}{\varepsilon_\infty}p^2 + (1-p^2)\right\}\frac{D_{a(b)}\varepsilon_\infty g_\parallel + g_\perp}{1-C_{g(b)}} \tag{10.38b}$$
$$C_{b(b)} = \frac{1+C_{g(b)}}{1-C_{g(b)}} \tag{10.38c}$$
$$C_{c(b)} = \frac{\Delta\chi_\mathrm{D}^0}{2\varepsilon_\infty}\frac{\varepsilon_\perp}{\varepsilon_b}\left\{\frac{\varepsilon_b}{\varepsilon_\infty}p^2 + (1-p^2)\right\}\frac{D_{a(b)}\varepsilon_\infty g_\parallel + g_\perp}{1-C_{h(b)}} \tag{10.38d}$$
$$C_{d(b)} = \frac{\frac{\Delta\chi_\mathrm{D}^0}{2\varepsilon_\infty}\left(\frac{\varepsilon_\infty}{\varepsilon_b}g_\perp - p^2\right)}{1-C_{h(b)}} \tag{10.38e}$$
$$C_{e(b)} = \frac{\Delta\chi_\mathrm{L}^0}{2\varepsilon_\infty}\frac{\varepsilon_\perp}{\varepsilon_b}\left\{\frac{\varepsilon_b}{\varepsilon_\infty}p^2 + (1-p^2)\right\}\frac{D_{a(b)}\varepsilon_\infty g_\parallel + g_\perp}{1-C_{i(b)}} \tag{10.38f}$$
$$C_{f(b)} = \frac{\frac{\Delta\chi_\mathrm{L}^0}{2\varepsilon_\infty}\left(\frac{\varepsilon_\infty}{\varepsilon_b}g_\perp - p^2\right)}{1-C_{i(b)}} \tag{10.38g}$$
$$C_{g(b)} = \frac{\chi^0}{2\varepsilon_\infty}\left[\frac{\varepsilon_\perp}{\varepsilon_b}g_\perp\left\{\frac{\varepsilon_b}{\varepsilon_\infty}p^2 + (1-p^2)\right\} - p^2\right] \tag{10.38h}$$

$$C_{h(b)} = \frac{\chi_{\mathrm{D}}^{0}}{2\varepsilon_{\infty}}\left[\frac{\varepsilon_{\perp}}{\varepsilon_{b}} g_{\perp}\left\{\frac{\varepsilon_{b}}{\varepsilon_{\infty}} p^{2} + (1-p^{2})\right\} - p^{2}\right] \tag{10.38i}$$

$$C_{i(b)} = \frac{\chi_{\mathrm{L}}^{0}}{2\varepsilon_{\infty}}\left[\frac{\varepsilon_{\perp}}{\varepsilon_{b}} g_{\perp}\left\{\frac{\varepsilon_{b}}{\varepsilon_{\infty}} p^{2} + (1-p^{2})\right\} - p^{2}\right] \tag{10.38j}$$

Thus, the coefficients in the case of Fig. 10.4 are derived by applying (10.36) and (10.37) to (10.38).

Finally, we study an example of a case where the medium boundary in the cell and the Faraday's law integration line (red line) do not intersect. This corresponds to the case shown in Fig. 10.5. In this case, since the integration line (red line) does not intersect the metal/background boundary in the cell, the effective dielectric constant $\varepsilon_{\perp}$ and the coefficients $c_{\perp}$ and $g_{\perp}$ are expressed by the following equations for cases Fig. 10.5a, c, e, and b, d, f, respectively:

$$\varepsilon_{\perp} = \varepsilon_{\infty}, \quad c_{\perp} = 0 \tag{10.39a}$$

$$\varepsilon_{\perp} = \varepsilon_{\mathrm{b}}, \quad g_{\perp} = 0 \tag{10.39b}$$

As in Figs. 10.3 and 10.4, when deriving the coefficients for cells (a), (c), and (e) in Fig. 11.5, (10.28) and (10.35a, 10.35b) are applied to (10.34). On the other hand, for cells (b), (d), and (f), the coefficients for each cell are derived by applying (10.37) to (10.38). The same procedure can be used to derive the coefficients for dispersive media of any shape other than spheres.

10.3.3 Analysis of Metallic Sphere

The DCP algorithm is used to analyze a metallic sphere model. First, the scattering cross section is evaluated. Au is used for the metallic sphere, and the model is treated as the Drude-Lorentz model to account for its dispersion. The background is air ($\varepsilon_{\mathrm{b}} = 1$) and the radius of the metallic sphere is chosen to be $a = 120\,\mathrm{nm}$. The E_x polarization with uniform amplitude in the x direction is excited in the $+z$ axis using the total field and scattered field domain decomposition (TF/SF) technique [8]. The spatial width is determined by the number of sample points N with a sphere diameter of $2a$ and $\Delta x = \Delta y = \Delta z = 2a/N$.

Figure 10.11 shows the wavelength characteristic of the scattering cross section. For comparison, the results obtained from the staircase approximation and the analytical result from the Mie theory [9] are also included. The DCP-FDTD method produces the result closer to that of the Mie theory than the staircase approximation. The DCP-FDTD method also gives similar accuracy to the staircase approximation, even when the number of sampling points is chosen to be about half that of the staircase approximation.

Next, we evaluate the scattering cross sections for a sample number of sphere diameters. Figure 10.12 shows the scattering cross sections for the number of

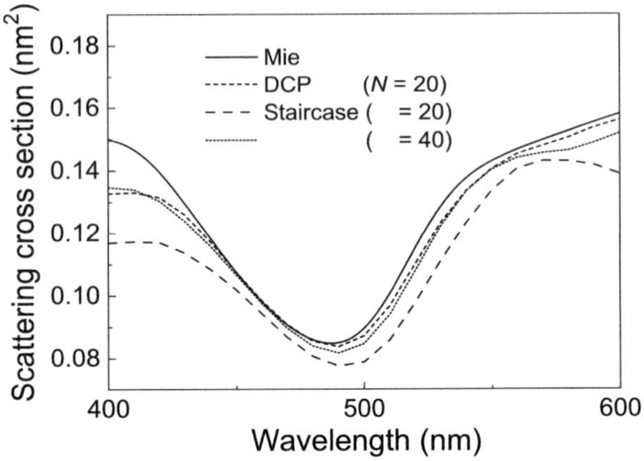

Fig. 10.11 Scattering cross-section versus wavelength

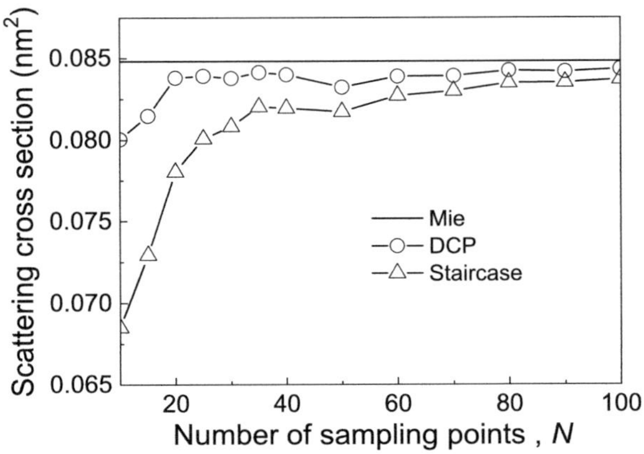

Fig. 10.12 Scattering cross-section as a function of the number of sampling points

sampling points in the diameters at a peak wavelength of $\lambda = 487\,\text{nm}$, and Fig. 10.13 shows the relative error to Mie theory. It can be seen that the result obtained from the DCP algorithm at $N = 20$ is closer to that of Mie theory than that obtained from the staircase approximation at $N = 100$, indicating a faster convergence of the numerical results.

Finally, we evaluate the field distribution (E_x component) in the x-z section near the metal sphere. The wavelength is chosen to be $\lambda = 487$ nm and the number of sampling points in the diameter is $N = 40$. Figure 10.14a shows the field distribution when the DCP-FDTD method is used, and (b) shows the field distribution when the staircase approximation is used. It can be seen that the DCP-FDTD method produces a relatively smooth field distribution along the boundary between the metal sphere and the air. On the other hand, when the staircase approximation is used, the field along the boundary is generally distorted.

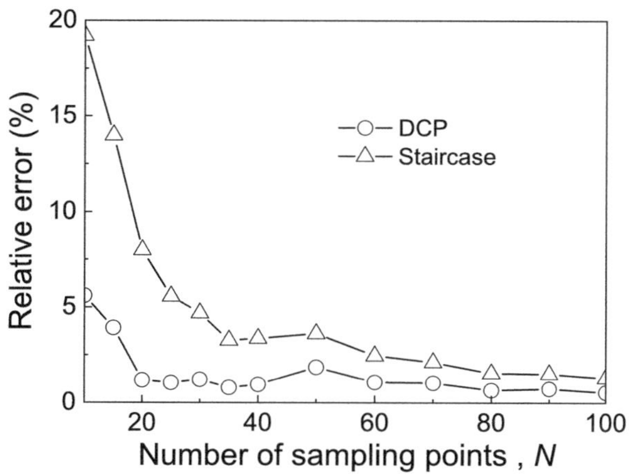

Fig. 10.13 Relative error as a function of the number of sampling points

10.4 Concluding Remarks

For the analysis of arbitrarily shaped dispersive media, we have reviewed the CP-EP technique for the analysis of arbitrarily shaped dielectric media. In particular, the CP-EP technique has been extended to the 3-D problem. For the 2-D problem, we have to deal with the six cases depending on the position of the dielectric in the rectangular cell. In contrast, for the 3-D problem, we have to handle fourteen cases depending on the position of the cubic cell. For the fourteen cases, the corresponding equations are derived and applied to the analysis of a dielectric sphere. Numerical results show that the convergence of the scattering cross section is much faster than that of the conventional staircase approximation. Next, we have extended the DCP technique to the 3-D problem using the similar methodology of the CP-EP technique. The fourteen cases have been studied for the analysis of the Drude-Lorentz model. It is found that even with a sampling number of $N = 20$, the DCP technique provides a result closer to the Mie result than the staircase approximation with $N = 100$. The field distribution obtained by the DCP technique is relatively smooth along the metal-dielectric boundary, while that of the staircase approximation is distorted not only along the boundary but also in the metal sphere.

Fig. 10.14 Field distributions

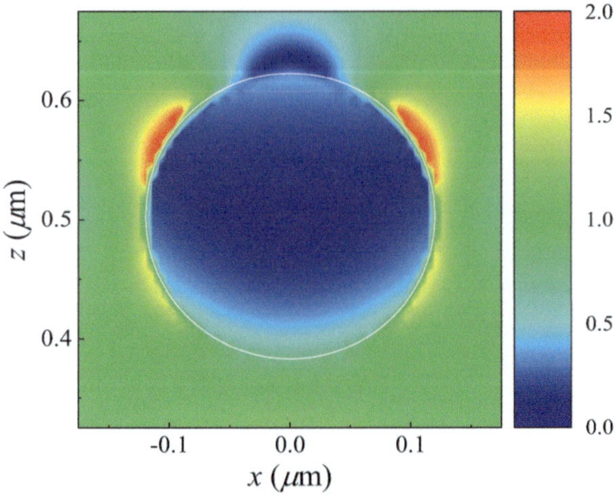

(a) DCP technique

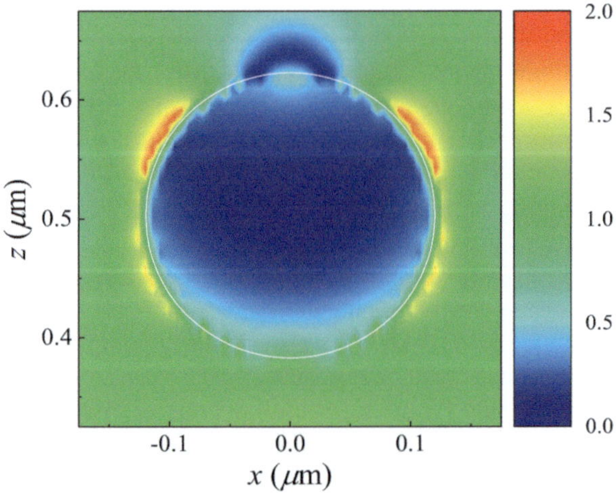

(b) Staircase

References

1. S. Dey, R. Mittra, A conformal finite-difference time-domain technique for modeling cylindrical dielectric resonators. IEEE Trans. Microw. Theory Tech. **47**(9), 1737–1739 (1999)
2. A. Mohammadi, H. Nadgaran, M. Agio, Contour-path effective permittivities for the two-dimensional finite-difference time-domain method. Opt. Exp. **13**(25), 10367–10381 (2005)
3. A. Mohammadi, M. Agio, Dispersive contour-path finite-difference time-domain algorithm for modelling surface plasmon polaritons at flat interfaces. Opt. Exp. **14**(23), 11330–11338 (2006)
4. A. Mohammadi, T. Jalali, M. Agio, Dispersive contour-path algorithm for the two-dimensional finite-difference time-domain method. Opt. Exp. **16**(10), 7397–7406 (2008)

5. J. Shibayama, K. Suzuki, J. Yamauchi, H. Nakano, Trapezoidal recursive convolution-based FDTD method for arbitrary-shaped dispersive materials. Electron. Lett. **54**(25), 1429–1430 (2018)
6. J. Shibayama, K. Suzuki, T. Iwamoto, J. Yamauchi, H. Nakano, Dispersive contour-path FDTD algorithm for the Drude-Lorentz model. IEEE Antennas Wirel. Propag. Lett. **19**(10), 1699–1703 (2020)
7. K. Takeya, T. Iwamoto, J. Shibayama, J. Yamauchi, H. Nakano, Extension of the CP-EP FDTD method to three-dimensional problems for the analysis of arbitrarily-shaped dielectric media. Microw. Opt. Technol. Lett. **65**(12), 3113–3119 (2023)
8. A. Taflove, S.C. Hagness, *Computational Electrodynamics: The Finite-Difference Time-Domain Method*, 3rd edn. (Artech House, Norwood, MA, 2005)
9. Q. Fu, W. Sun, Mie theory for light scattering by a spherical particle in an absorbing medium. App. Opt. **40**(9), 1354–1361 (2001)
10. J. Shibayama, R. Ando, A. Nomura, J. Yamauchi, H. Nakano, Simple trapezoidal recursive convolution technique for the frequency-dependent FDTD analysis of a Drude-Lorentz model. IEEE Photonics Technol. Lett. **21**(2), 100–102 (2008)

Chapter 11
Unidirectional Excitation of Plasmonic Grating

Abstract This chapter presents the analysis of a plasmonic grating coupler using the FDTD method. For the conventional coupler, the shape of the dielectric grating on the metal substrate is rectangular. In this case, the field coupled to the grating propagates in both directions on the metal surface at normal incidence. When this coupler is used for an optical link, half of the coupled power is wasted. At oblique incidence, the coupled field propagates unidirectionally on the metal surface. However, it is relatively difficult to precisely control the angle of incidence. To overcome this problem, a trapezoidal dielectric grating coupler is proposed for unidirectional propagation at normal incidence. Based on Snell's law, the trapezoidal grating is designed so that the metal substrate receives oblique incident light through the trapezoidal grating even under normal incidence. It is found that the coupling light in the grating propagates unidirectionally on the metal surface even at normal incidence. It should be noted that the slanted shape of the trapezoidal grating is relatively difficult to fabricate. Therefore, a grating coupler with stepped dielectrics is also investigated. The field propagates unidirectionally on the metal surface at normal incidence. Finally, the possibility of reducing the reflectivity by adding an anti-reflection (AR) coating is considered. It is shown that the reflectivity is significantly suppressed at 1.55 μm and the field can propagate in one direction.

11.1 Introduction

Surface plasmon polaritons (SPPs) are known to propagate along the boundary between metal and dielectric materials in the lightwave band [1]. SPPs can be excited in plasmonic waveguides using a grating coupler [2]. In [3, 4], optical links were investigated using two grating couplers loaded with rectangular dielectrics through a waveguide. It was shown that vertically incident light can be coupled as an SPP to the input grating. The coupled field propagates through the plasmonic waveguide, and can then be extracted as output light from the other grating. However, as shown in the literature [3], the coupled field at the grating propagates in both directions on the waveguide and only half of the coupling power can be utilized.

J. Shibayama, *FDTD Analysis of Plasmonic Devices*, Springer Series in Optical Sciences 254, https://doi.org/10.1007/978-981-96-8130-3_11

Incidentally, we have investigated InSb-based grating couplers in the terahertz band in Chap. 9. InSb-coated dielectric cylinders were arrayed on a slab substrate and the transmission characteristics and field behavior were investigated [5]. It was found that the coupling of the incident wave to the substrate results in multiple transmission dips and that the surface plasmon resonance occurs at specific wavelengths. It was also shown that the coupled fields propagate unidirectionally through the substrate at oblique incidence. Furthermore, InSb sphere arrays [6] were analyzed in the three-dimensional structures in Sect. 9.4. The Poynting vectors at normal and oblique incidence were evaluated, and the direction of field propagation was discussed.

In this section, we analyze a plasmonic grating coupler in the lightwave band by the FDTD method and investigate the realization of unidirectional field propagation at normal incidence [7]. First, we evaluate the reflection characteristics, the behavior of the field at the dip wavelength, and the propagation direction of the field in a grating coupler with a rectangular dielectric loaded on a metal substrate. It is pointed out that the field propagates in both directions on the metal surface at normal incidence, while the field propagates unidirectionally on the metal surface at oblique incidence. Next, we consider the case where the dielectric shape is changed to a trapezoidal shape in order to achieve unidirectional field propagation at normal incidence. It is found that the field propagates unidirectionally on the metal surface despite the normal incidence. Furthermore, the period length of the grating is varied to adjust the coupling wavelength to a practical value of 1.55 μm.

It should be noted that the slanted shape of the trapezoidal grating is relatively difficult to fabricate. Therefore, we change the shape of a grating to a stepped dielectric for ease of fabrication. The field is shown to propagate unidirectionally on the metal surface at normal incidence. Finally, the reduction of reflectivity by adding an anti-reflection (AR) coating is considered. It is shown that at 1.55 μm the reflectivity is significantly suppressed and the field can propagate in one direction.

11.2 Plasmonic Grating Coupler Loaded with Rectangular Dielectrics

The case of rectangular dielectrics loaded on a metal substrate is analyzed for both normal and oblique incidence. The structure of the coupler to be analyzed is shown in Fig. 11.1. Only one period ($\Lambda = 1.2$ μm) of the infinite period structure for the x direction is shown in Fig. 11.1b. The material of the rectangular dielectric is SiO_2 and its refractive index is chosen to be 1.444, the height to be $h = 0.775$ μm and the width to be $w = 0.72$ μm. Ag is chosen as the metal which is thick enough, and the relative permittivity is expressed by the following Drude model [8].

$$\varepsilon_r(\omega) = \varepsilon_\infty + \frac{\omega_{\mathrm{D}}^2}{\omega\,(j\nu_{\mathrm{D}} - \omega)} \tag{11.1}$$

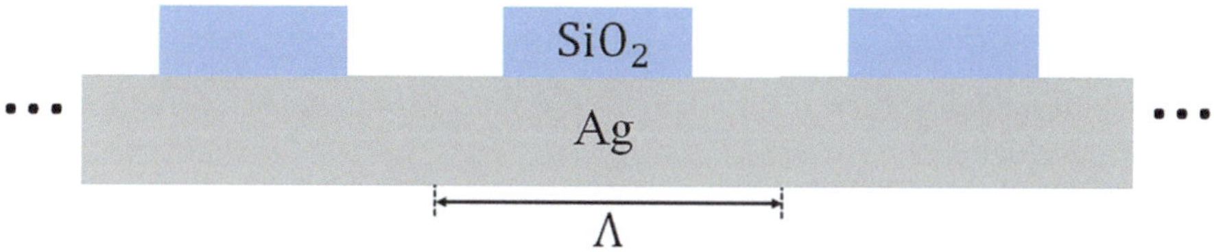

(a) Grating coupler with rectangular dielectrics

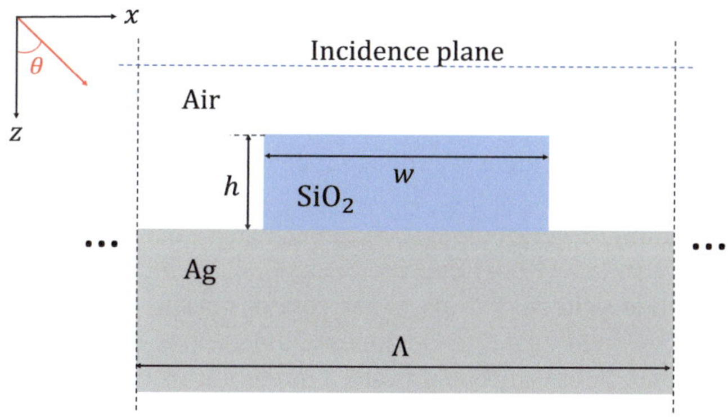

(b) Unit cell

Fig. 11.1 Grating structure. Copyright © 2025 IEICE from [7]

where ε_∞ is the dielectric constant of the material at infinite frequency, ω is the angular frequency, ω_D is the electron plasma frequency, ν_D is the effective electron collision frequency. This Drude model is incorporated into the FDTD method based on the trapezoidal recursive convolution presented in Chap. 3. The periodic boundary condition [9] is applied to the analysis, and only one period is treated. The incident wave is excited with $E_x(H_y)$ polarization with uniform amplitude in the transverse direction, tilted θ from the z axis. The spatial sampling widths are chosen to be $\Delta x = \Delta z = 5$ nm.

Figure 11.2 shows the reflection characteristics for $\theta = 0$ and $20°$ incident angles. It can be seen that dips occur at both angles of incidence. When the incident wave satisfies the phase matching conditions [2, 5], the field at the metal surface is coupled as SPP at the dip wavelength. The dip wavelength is shown to shift to the long wavelength side when θ is increased. The dip wavelength is 1.55 μm for $\theta = 0°$ and 1.88 μm for $\theta = 20°$. The reflectance is 0.42 and 0.44 respectively.

Figure 11.3 shows the field distribution ($|H_y|$ components) at the dip wavelength, where (a) is for $\theta = 0°$ and (b) is for $\theta = 20°$. It is normalized to the maximum amplitude. It can be seen that SPPs are excited on the metal surface at both angles of

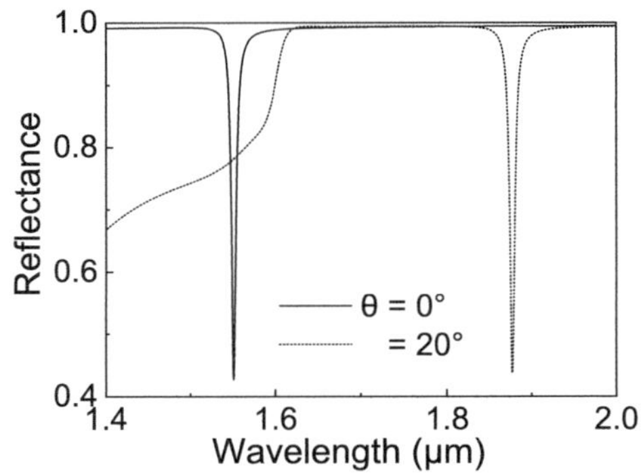

Fig. 11.2 Reflection characteristics for rectangular dielectrics.

incidence. In addition, standing waves due to reflected waves can be observed above the gratings.

Next, the direction of field propagation at each angle of incidence is evaluated. Figure 11.4a and b show the Poynting vectors at $\theta = 0°$ and $\theta = 20°$, respectively, with the maximum amplitude of 0 dB. In the case of normal incidence in (a), it can be confirmed that the field rotates in the opposite direction around the grating center. This implies that the field distributed in the x direction in Fig. 11.3a is a standing wave generated by the SPP propagating in the $\pm x$ directions. On the other hand, in the case of the oblique incidence in (b), it can be seen that the field propagates unidirectionally in the $-x$ direction on the metal surface. It is interesting to note that the field in the metal propagates in the $+x$ direction as indicated by the blue arrow. This may be due to the negative dielectric constant of the metal.

As a result, the field propagates in both directions on the metal surface at normal incidence, and only half of the coupled power can be utilized. In contrast, when the field is injected at oblique incidence, the field propagates in one direction and more of the coupled power can be utilized.

11.3 Plasmonic Grating Coupler Loaded with Trapezoidal Dielectrics

In the previous chapter, we showed that oblique incidence of the field can realize unidirectional propagation of a coupled field on a metal surface. However, it is generally difficult to control the angle of incidence with high precision. To enable unidirectional propagation of the field even at normal incidence, we consider the case where the shape of the dielectric is changed to a trapezoidal shape.

To do this, we here estimate the angle of incidence of the field on the metal surface inside the dielectric for rectangular dielectrics. At an angle of incidence of $\theta = 20°$, the field enters the metal surface at approximately 13.7° according to Snell's law. This angle is used in the following grating design.

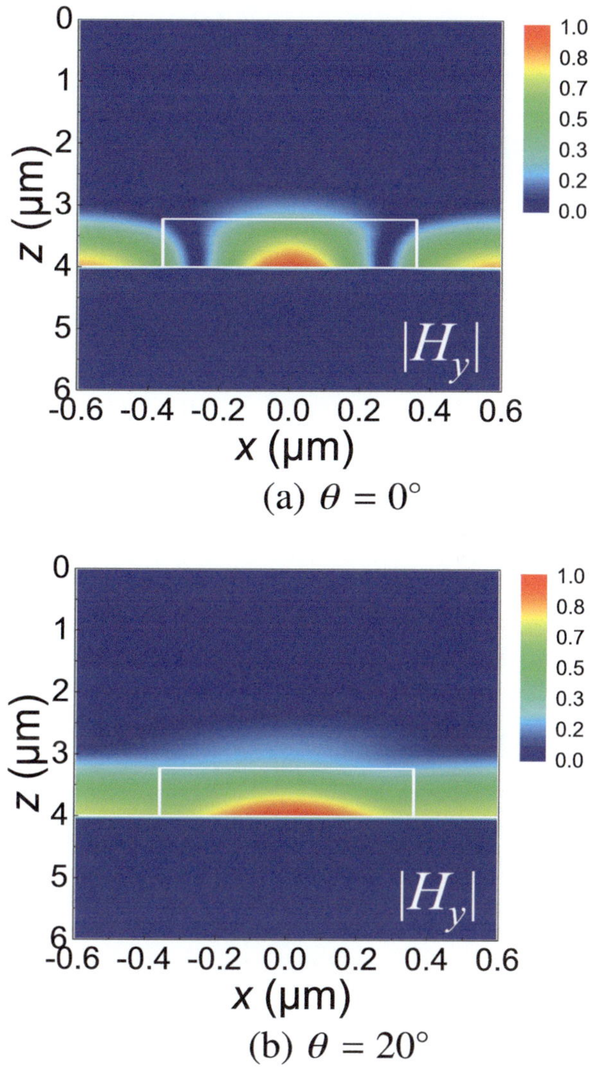

Fig. 11.3 Field distribution for rectangular dielectrics. Copyright © 2025 IEICE from [7]

The proposed structure of the coupler is shown in Fig. 11.5. The shape of the dielectric is chosen to be trapezoidal and the length of the short side is set to be $h_1 = 0.4$ μm. The period length is set to be $\Lambda = 1.2$ μm as in Fig. 11.1. We set the angle θ_1 and the length of the long side h_2 as shown in Fig. 11.6 so that the angle of refraction when the light is normally incident on the dielectric is equal to the condition when the rectangular dielectric is loaded in the previous section. The relationship between the angle of incidence θ_1 on the trapezoidal dielectric, the angle of incidence on the metal surface θ_2, and the refraction angle θ_3 is given by

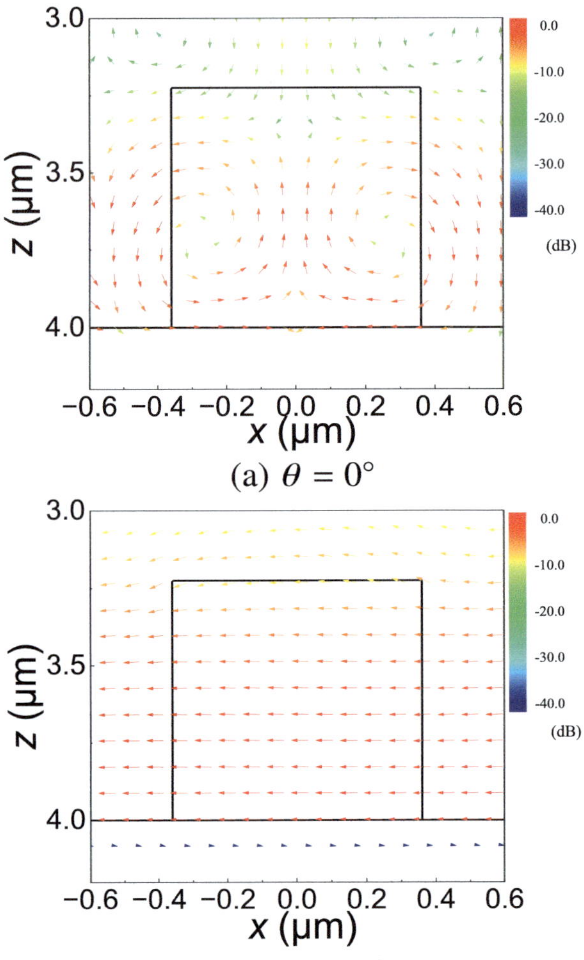

Fig. 11.4 Poynting vector for rectangular dielectrics. Copyright © 2025 IEICE from [7]

$$\theta_1 = \theta_2 + \theta_3 \tag{11.2}$$

Applying Snell's law to the relation in Fig. 11.6 leads to

$$n_{\mathrm{air}} \cdot \sin\theta_1 = n_{\mathrm{SiO_2}} \cdot \sin\theta_3 \tag{11.3}$$

then $\theta_1 \simeq 40.3°$ is derived with $\theta_2 \simeq 13.7°$. As a result, we obtain $h_2 \simeq 1.42\ \mu\mathrm{m}$ for the long side of the trapezoidal dielectric.

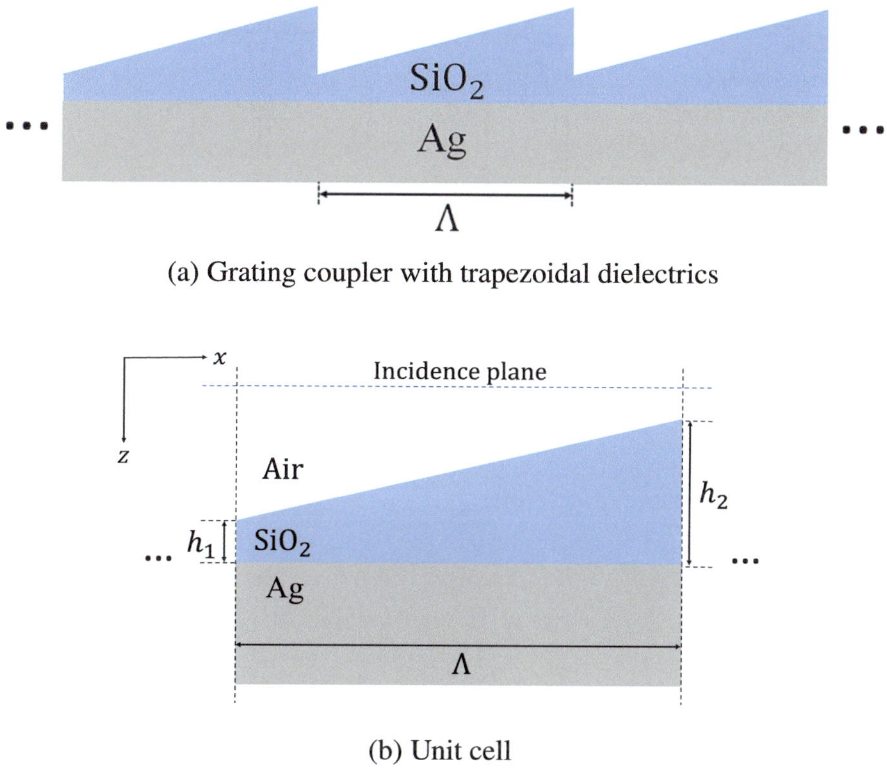

Fig. 11.5 Grating structure. Copyright © 2025 IEICE from [7]

Fig. 11.6 Design of the trapezoidal dielectric. Copyright © 2025 IEICE from [7]

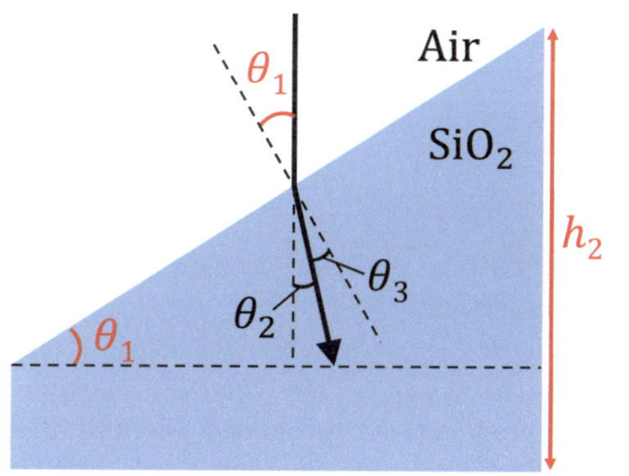

Figure 11.7 shows the reflection characteristics at normal incidence for the proposed trapezoidal grating. It can be seen that multiple reflection dips occur. This may be due to the fact that the trapezoidal dielectric satisfies multiple phase-matching conditions because of its continuously changing structure. The dip wavelengths are 1.68

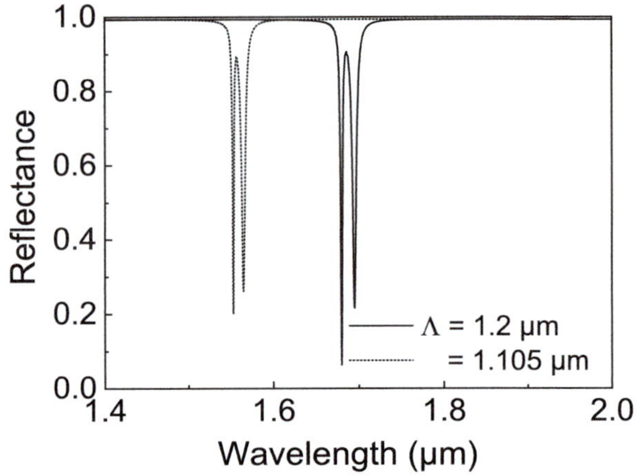

Fig. 11.7 Reflection characteristics for trapezoidal dielectrics. Copyright © 2025 IEICE from [7]

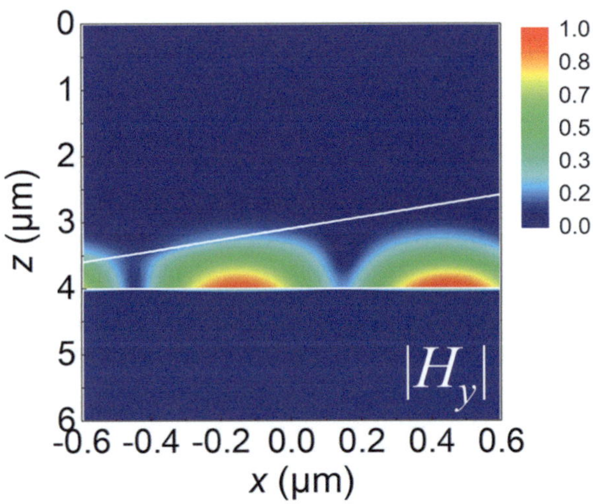

Fig. 11.8 Field distribution for trapezoidal dielectrics. Copyright © 2025 IEICE from [7]

and 1.69 μm. At these wavelengths, the incident field is coupled to the SPP on the metal surface.

Figure 11.8 shows the field distribution ($|H_y|$ component) at a dip wavelength of 1.68 μm. It can be seen that SPP is excited on the metal surface. The direction of field propagation is evaluated. Figure 11.9 shows the Poynting vector. It can be seen that the field propagates unidirectionally in the $-x$ direction on the metal surface even at normal incidence. Therefore, the field on the metal surface in Fig. 11.8 is considered to be a localized SPP rather than a standing wave. Although not shown, the field also propagates unidirectionally in the $-x$ direction at a dip wavelength of 1.69 μm.

The above shows that the trapezoidal shape of the dielectric allows the field to propagate in one direction even at normal incidence and to utilize a large amount of

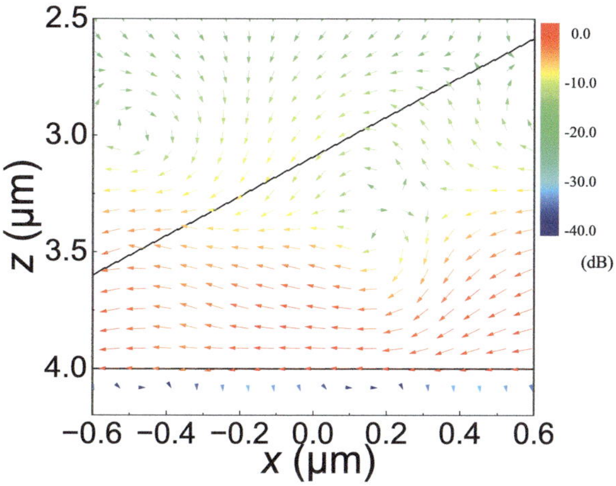

Fig. 11.9 Poynting vector for trapezoidal dielectrics. Copyright © 2025 IEICE from [7]

the coupled power. However, the coupling wavelength of the field is not practical. Therefore, the period length of the grating coupler is varied to obtain a coupling wavelength at a practical communication wavelength of 1.55 μm. Figure 11.7 also shows the reflection characteristics when the period length is chosen to be $\Lambda = 1.105$ μm. Again, multiple dips occur, and the field is coupled to the metal surface at the dip wavelength. The dip wavelengths for $\Lambda = 1.105$ μm are 1.55 and 1.56 μm. Although not shown, the field propagates unidirectionally in the $-x$ direction on the metal surface, similar to the result in Fig. 11.9.

11.4 Modification of the Grating Structure

By using trapezoidal dielectrics, as discussed above, the field propagates in one direction even at normal incidence, leading to the possibility of using a large amount of the coupled power. However, it is generally difficult to accurately fabricate the slope of trapezoidal dielectrics. Here, we investigate a grating with stepped dielectrics, which are easier to fabricate than the trapezoidal dielectrics.

Figure 11.10 shows the grating coupler with stepped dielectrics. The number of the steps is set to 5, and their heights and widths are chosen to be $h_3 = 0.20$ μm and $w = 0.24$ μm, respectively. Figure 11.11 shows the reflection characteristic. The solid line in the figure shows the result of the trapezoidal dielectrics, and the dotted line shows the that of the stepped dielectrics. It can be seen that the stepped dielectric also has two dips. The dip wavelengths are 1.663 and 1.687 μm, and the reflectivities are 0.07 and 0.08, respectively. The H_y field distribution and the Poynting vector at 1.663 μm are shown in Figs. 11.12 and 11.13, respectively. The field distribution shows that the SPPs are excited on the metal surface. The Poynting vector shows that the field propagates in one direction, as with the trapezoidal dielectrics.

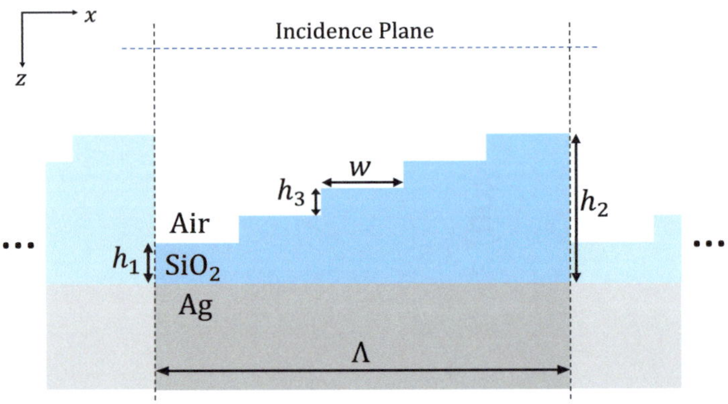

Fig. 11.10 Grating coupler with stepped dielectrics

Fig. 11.11 Reflection characteristics for stepped dielectrics

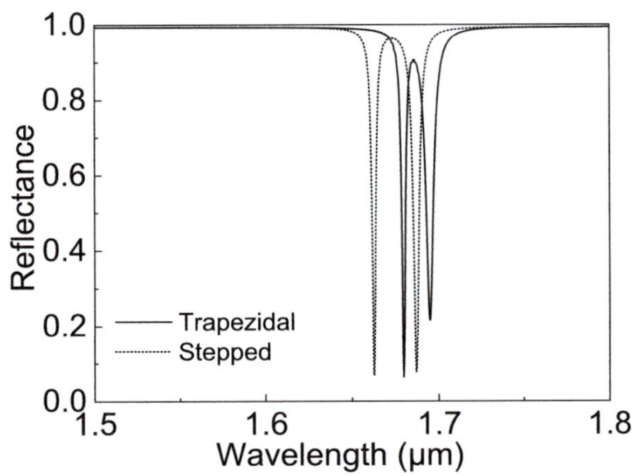

The final consideration will be the use of an anti-reflection (AR) coating to further reduce the reflectivity of the grating coupler, the configuration of which is shown in Fig. 11.14. The refractive index of the AR coating is set to 1.20, which is the square root of the product of the refractive index of silica and air, and the thickness is set to $h_{\mathrm{AR}} = 0.39$ μm, which is one quarter wavelength of the desired wavelength. To adjust the coupled wavelength, we set the grating period is $\Gamma = 1.09$ μm. Figure 11.15 shows the reflection characteristics, in which the result without the AR coating is included for $\Gamma = 1.2$ μm. This figure shows that the dip occurs at a practical wavelength of 1.55 μm by adjusting the period length Γ from 1.2 to 1.09 μm. The reflectivity at the dip wavelength is 0.07 for the solid line and 0.001 for the dotted line, which shows that the reflectivity is suppressed by loading the AR coating. The H_y field distribution and the Poynting vector are shown in Figs. 11.16 and 11.17, respectively. The field on the metal surface propagates in one-direction, even when the reflectivity is suppressed by AR coatings.

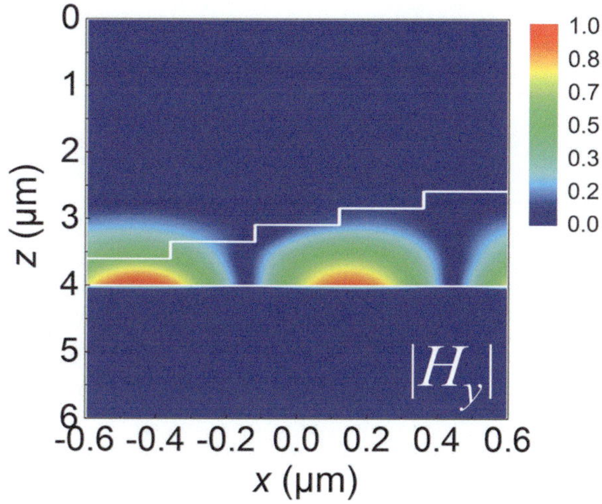

Fig. 11.12 Field distribution for stepped dielectrics

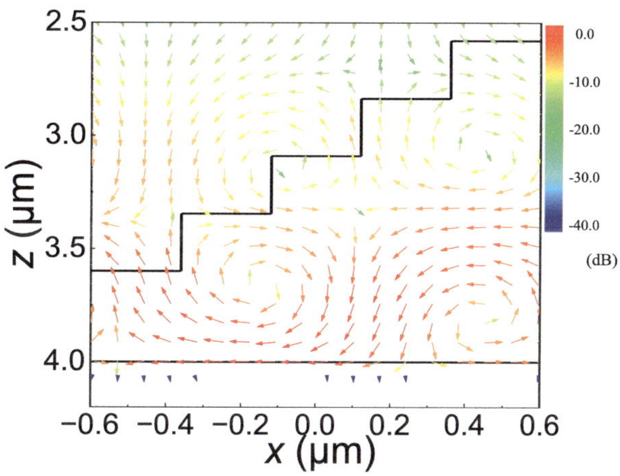

Fig. 11.13 Poynting vector for stepped dielectrics

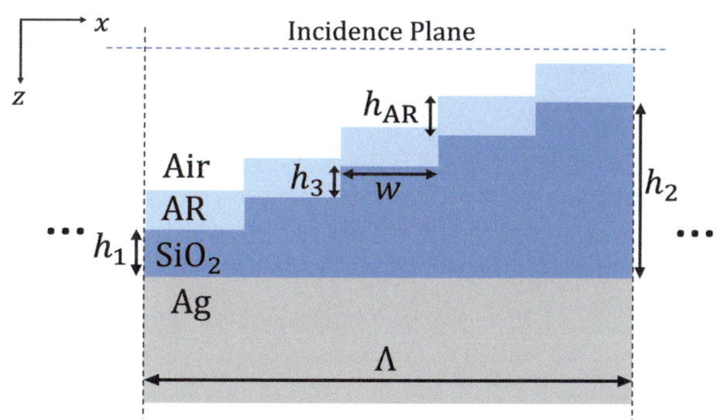

Fig. 11.14 Grating coupler with stepped dielectrics and AR coating

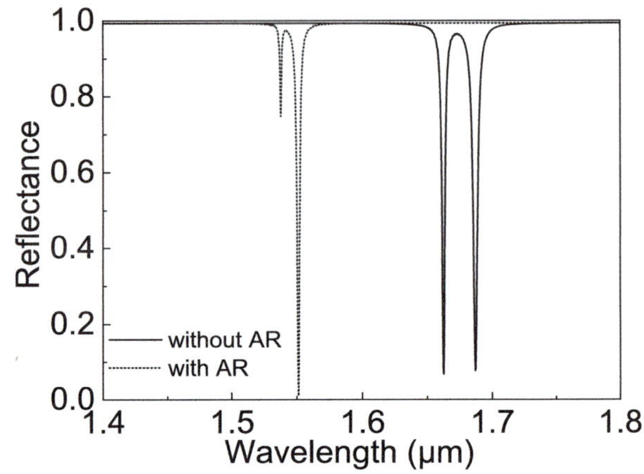

Fig. 11.15 Reflection characteristics for stepped dielectrics and AR coating

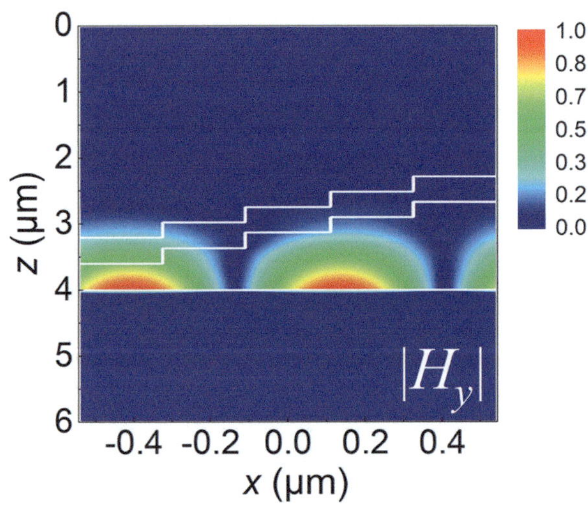

Fig. 11.16 Field distribution for stepped dielectrics and AR coating

11.5 Concluding Remarks

We have analyzed plasmonic grating couplers considering the shape of the dielectric materials using the FDTD method at optical frequencies for unidirectional excitation of normal incident light. First, a grating coupler with rectangular dielectrics has been investigated. It is shown that the unidirectional excitation is realized when the incident light is excited at oblique incidence. Next, a grating coupler with trapezoidal dielectrics was analyzed for one-way excitation even at normal incidence. The coupled field on the grating propagates unidirectionally on the metal surface at normal incidence. A grating coupler with stepped dielectrics has been investigated. It is confirmed that the field propagates in one direction on the metal surface at normal incidence. Final consideration has been given to reducing the reflectivity adding an

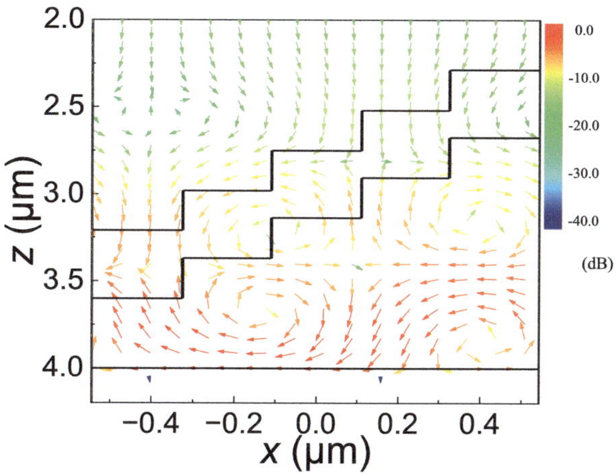

Fig. 11.17 Poynting vector for stepped dielectrics with AR coating

AR coating. It is found that the reflectivity is significantly suppressed at 1.55 μm, and the field can propagate unidirectionally.

References

1. M.L. Brongersma, P.G. Kik (ed.), *Surface Plasmon Nanophotonics* (Springer, Dordrecht, 2007)
2. S. Ura, S. Murata, Y. Awatsuji, K. Kintaka, Design of resonance grating coupler. Opt. Exp. **16**(16), 12207–12213 (2008)
3. M.G. Nielsen, J.-C. Weeber, K. Hassan, J. Fatome, C. Finot, S. Kaya, L. Markey, O. Albrektsen, S.I. Bozhevolnyi, G. Millot, A. Dereux, Grating couplers for fiber-to-fiber characterizations of stand-alone dielectric loaded surface plasmon waveguide components. J. Light. Technol. **30**(19), 3118–3125 (2012)
4. D. Wu, K. Hamashima, S. Inoue, S. Ohnuki, Design and characteristic verification of plasmonic waveguides using finite-difference complex-frequency-domain method. IEICE Trans. C **J103-C**(2), 69–77 (2020) (in Japanese)
5. J. Shibayama, S. Takahashi, J. Yamauchi, H. Nakano, Fundamental investigation of a grating consisting of InSb-coated dielectric cylinders on a substrate in the THz regime. IEICE Trans. Electron. **E103-C**(11), 567–574 (2020)
6. J. Shibayama, T. Kuroda, J. Yamauchi, H. Nakano, Analysis of an InSb sphere array on a dielectric substrate in the THz regime. IEICE Trans. Electron. **E105-C**(4), (2022)
7. H. Nakasaka, K. Takeya, J. Shibayama, FDTD analysis of a plasmonic grating coupler using rectangular or trapezoidal dielectric grating. IEICE Trans. Electron. **J108-C**, 40–50 (2024) (Japanese Edition)
8. P.B. Johnson, R.W. Christy, Optical constants of the noble metals. Phys. Rev. B **6**(12), 4370–4379 (1972)
9. A. Taflove, S.C. Hagness, *Computational Electrodynamics: The Finite-Difference Time-Domain Method*, 3rd edn. (Artech House, Norwood, MA, 2005)

Index

J. Shibayama, *FDTD Analysis of Plasmonic Devices*, Springer Series in Optical Sciences 254, https://doi.org/10.1007/978-981-96-8130-3

MIX
Papier aus verantwortungsvollen Quellen
Paper from responsible sources
FSC® C105338

If you have any concerns about our products,
you can contact us on
ProductSafety@springernature.com

In case Publisher is established outside the EU,
the EU authorized representative is:
Springer Nature Customer Service Center GmbH
Europaplatz 3, 69115 Heidelberg, Germany

Printed by Libri Plureos GmbH
in Hamburg, Germany